Design of Cities and Buildings - Sustainability and Resilience in the Built Environment

Edited by Samad Sepasgozar, Sara Shirowzhan, Sharifeh Sargolzae and José David Bienvenido-Huertas

Published in London, United Kingdom

IntechOpen

Supporting open minds since 2005

Design of Cities and Buildings - Sustainability and Resilience in the Built Environment
http://dx.doi.org/10.5772/intechopen.87647
Edited by Samad Sepasgozar, Sara Shirowzhan, Sharifeh Sargolzae and José David Bienvenido-Huertas

Contributors
Fermín Alí Cruz-Muñoz, Sara Velasco, joaquim Maloa, Hatem Ibrahim, Tsutomu Tsuboi, Nilüfer Kart Aktaş, Nazlı Yıldız Dönmez, Tao Ma, Xiaolei Wang, Hao Wang, Nairong Tan, Zhou Mingxi, Janmejoy Gupta, Prashanti Rao, Anil Kar, André De Naeyer, Wael Zatar, Hai Nguyen, Samad M.E. Sepasgozar, José David Bienvenido-Huertas, Sara Shirowzhan, Sharifeh Sargolzae

First published in London, United Kingdom, 2021 by IntechOpen
IntechOpen is the global imprint of INTECHOPEN LIMITED, registered in England and Wales, registration number: 11086078, 5 Princes Gate Court, London, SW7 2QJ, United Kingdom
Printed in Croatia

British Library Cataloguing-in-Publication Data
A catalogue record for this book is available from the British Library

Additional hard and PDF copies can be obtained from orders@intechopen.com

Design of Cities and Buildings - Sustainability and Resilience in the Built Environment
Edited by Samad Sepasgozar, Sara Shirowzhan, Sharifeh Sargolzae and José David Bienvenido-Huertas
p. cm.
Print ISBN 978-1-78985-149-6
Online ISBN 978-1-78985-150-2
eBook (PDF) ISBN 978-1-78985-669-9

For EU product safety concerns:
IN TECH d.o.o., Prolaz Marije Krucifikse Kozulić 3, 51000 Rijeka, Croatia, info@intechopen.com or visit our website at intechopen.com.

Meet the editors

Dr. Samad Sepasgzoar is a senior lecturer and the co-convenor for the smart city and infrastructure cluster. He is also an associate editor, editorial board member, guest editor, or reviewer for more than thirty-five peer-reviewed journals. He has published 137 research articles. His research is cross-interdisciplinary, using mixed statistical, experimentation, and computational methods to analyse complex issues of digital technology development and implementation processes. Due to the quality of his research projects, Dr. Sepasgzoar has been recognised internationally by different industrial bodies with awards such as 'Best Paper'. He was also a finalist for the 'Australian Construction Awards'.

Dr. José David Bienvenido-Huertas, Ph.D., is an architect and researcher for the Department of Building Construction II at the University of Seville, Spain. He is an active member of the Research Group TEP970: Technological Innovation, 3D Modeling Systems and Energy Diagnosis in Heritage and Building at the University of Seville. His areas of expertise include climate change in the building sector, adaptive thermal comfort, heat transfer, fuel poverty, energy conservation measures, and the design of nearly zero-energy buildings. He is an author of more than fifty research papers. Moreover, Dr. Huertas is a recognized reviewer of various international indexed journals and international research projects.

Dr. Sara Shirowzhan is a lecturer and Academic Advisor for the City Analytics Program, School of Built Environment, University of New South Wales (UNSW), Sydney, Australia. Her areas of research include sensing technologies, advanced GIS, BIM, digital twins, and artificial intelligence. She teaches and supervises students on city analytics, GIS, construction informatics, and BIM topics at undergraduate and postgraduate levels at UNSW. Dr. Shirowzhan is currently an editorial board member for the journal *Sustainability* and *Advances in Civil Engineering*. She is also a topic board member of the *ISPRS International Journal of Geo-Information*. Dr. Shirowzhan obtained her Ph.D. in Geomatics Engineering from the School of Civil and Environmental Engineering, UNSW.

Dr. Sharifeh Sargolzae has a Ph.D. in Urban and Regional Planning and is a lecturer in the Department of Urban Planning, University of Zabol, Iran. Her expertise covers technology adoption and acceptance in urban management fields, especially in new technologies applied in municipalities. Her recent research project focused on adopting new technologies to improve urban areas' quality of life at both individual and organizational levels. Other experiences include working on issues such as smart cities, urban planning, disaster management, and emergency accommodation.

Contents

Preface

The impacts from building construction and use contribute to extreme environmental degradation. On one hand, constructions generate an environmental impact through the resources used and the greenhouse gas emissions generated during the construction process. On the other hand, buildings, once constructed and put into use, consume large amounts of energy. For this reason, it is necessary to have efficient designs to mitigate the environmental impact of buildings during their different phases, with the goal of achieving a low-carbon building stock by 2050.

There are many reasons and evidence showing that the design and construction sectors have a significant role in pollution and emissions. Thus, sustainability in these sectors is the core of many investigations and scholarly works. Many statistics show that these sectors account for more than 40 percent of emissions globally.

This book discusses the most appropriate architectural design strategies that guarantee the adequate environmental performance of buildings during the design and construction phase as well as the use phase. It collects and presents the outcomes of investigations and discussions about sustainable concepts and urban planning and development. Sustainability is an interdisciplinary topic requiring innovation and creativity. This means innovative works in different disciplines should be considered to improve the quality of the built environment in terms of economic, environmental, and social attributes. The book begins the discussion with a snapshot of trending topics and how intelligence systems may revive businesses and industries related to smart cities and construction. Overall, it focuses on relevant issues related to the production of construction materials, lifecycle analysis, sustainable constructions, efficient designs, nearly zero-energy buildings, and the sociocultural integration aspects of new architectural designs in urban settings.

Following the Introduction, the second section includes chapters focusing on durable concrete, climate resilience, sustainable building, energy-efficient design, and innovative sustainable construction topics. Concrete is one of the main materials used in construction, but it is also a source of pollution. Thus, the durability of this material becomes important for improving the sustainability of projects.

The third and last section focuses on urban development topics such as traffic flow analysis, architecture, urban resilience, and human needs. Traffic management will affect the structure of cities including the number of required infrastructure projects, which can affect a city's sustainability and resilience. An efficient management system will consider adapting a city's current infrastructure as well as add to it with, for example, reserve lanes or lanes for cyclists. Another important possibility of traffic management is to provide rerouting options to citizens in case of traffic jams, which may save fuel and energy.

Samad Sepasgozar and Sara Shirowzhan
The School of Built Environment,
University of New South Wales,
Sydney, Australia

José David Bienvenido-Huertas
Department of Building Construction II,
University of Seville,
Sevilla, Spain

Sharifeh Sargolzae
Department of Urban Planning,
Zabol University,
Zabol, Sistan and Baluchestan, Iran

Section 1

Introduction

Chapter 1

Introductory Chapter: Intelligence, Sustainable and Post-COVID-19 Resilience Built Environment: An Agenda for Future

Samad Sepasgozar, José David Bienvenido-Huertas, Sara Shirowzhan and Sharifeh Sargolzae

1. Introduction

Digital communication and information technologies were the core of businesses during the COVID-19 pandemic. Many businesses and tasks have been done remotely, such as business meetings, education, and e-commerce. There are many reports that show technology played an essential role in response to the pandemic and is predicted to be the core strategy for long-term resilience.

The recent reports show that the use of digital technologies and further technical advances occurred due to COVID-19 in different sectors, including infrastructure and transportation [1]. The initial readiness for adopting the Fourth Industrial Revolution concept [2, 3] in the architecture, engineering, and construction (AEC) sector was helpful for industry resilience during COVID-19. The current reports show that the adoption process of Industry 4.0 technologies will be accelerated due to the community's higher demand in terms of working remotely with less human close interactions [1]. The current investigations show that site managers and operational teams in the construction, transportation, and infrastructure sectors need to monitor the progress of tasks and keep track. In addition, they should assist their team members and ensure they will achieve the project objectives, including time, cost, and quality. However, the investigations show that the lockdown caused many difficulties to managers' tasks and attending the construction site or infrastructures [1]. However, designers using digital technologies such as Building Information Modeling (BIM) [4–6] or Geographic Information System (GIS) [7–9] experienced much less difficulties. This chapter aims to provide an insight into the recent literature of COVID-19 in the built environment. Then, suggestions for future studies are presented.

2. Scientometric analysis of COVID-19 in the built environment

The scientometric analysis technique is adopted to identify key knowledge themes and map them to identity overlaps, gaps, emerging topics and monitor the growth of the literature with the included patterns. This is a recommended approach since many studies in the field used scientometric analysis techniques in different contexts such as lean construction [10], additive manufacturing [11], carbon emission [12], smart home [13], delay analysis [14].

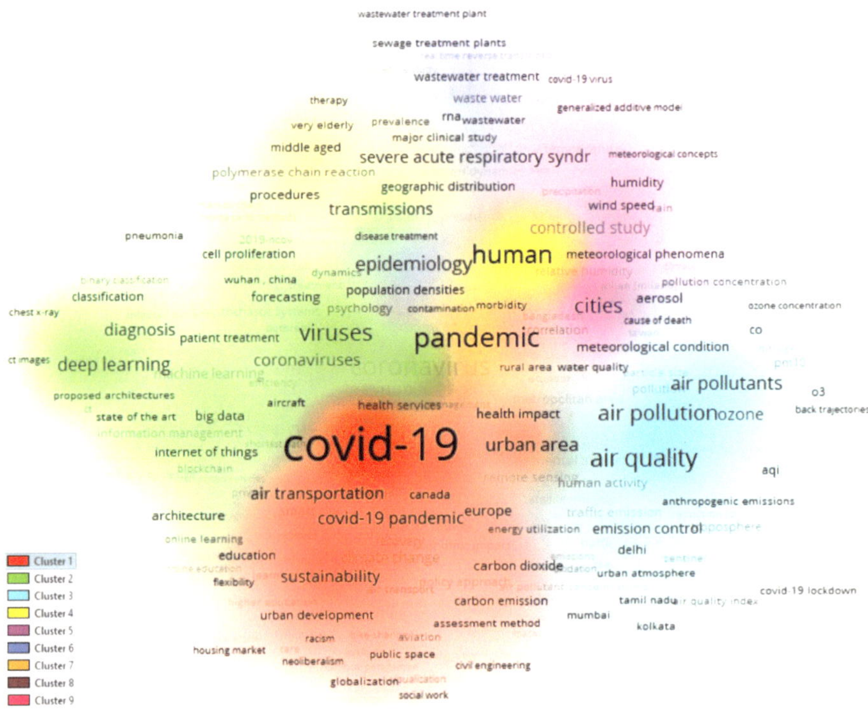

Figure 1.
Co-occurrence analytical map of keywords created on the COVID-19 literature (see the first set of strings in appendix). A total of 1308 keywords out of the sample of 9460 keywords are shown since they meet the threshold with the minimum co-occurrence of 3.

Figure 1 shows nine clusters of keywords in the recent publications that appeared on the Scopus database, including 2,185 journal articles. The main cluster has a strong co-occurrence with COVID-19 and refers to different themes such as sustainability, urban development, and public spaces. Other important clusters refer to humans, cities, and deep learning. The air quality, humidity, wind speed, and other environmental attributes received significant attention. However, some other themes, such as the role of digital technologies for post-pandemic and improving the resilience of the country and cities, are slowly developing.

3. An agenda for future studies

This chapter reviewed the current literature of COVID-19 and post-COVID-19 to understand the literature's themes and directions. However, more detailed investigations are required to use different databases and identify the suggested solutions for post-COVID-19 in the businesses relevant to the built environment, including transportation and construction. The impact of the pandemic on the construction industry has been evaluated in different contexts and countries such as Oman [15], Jordan [16], the United Kingdom [1, 17], Seri Lanka [18], South Africa [19], Japan [20] and Argentina [21]. However, most of the publications are affiliated with authors from developed countries such as the United States of America, the United Kingdom, Canada, Italy, Germany, and Spain, including a few developing countries (see **Figure 2**).

Besides the growing literature of 'impact' evaluations, there will be many industry and government reports available providing evidence of affected firms and

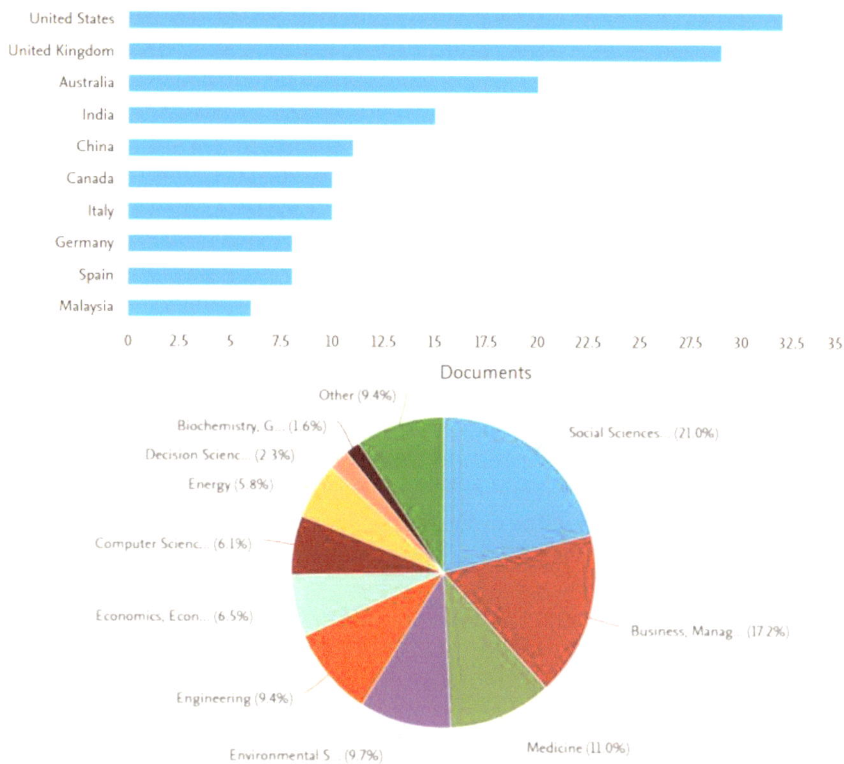

Figure 2.
The selected literature analysis is based on countries and subject areas (see the second set of strings in appendix).

people. However, some studies suggest that the impact analysis investigations are not enough yet, and some other affected areas should be evaluated. For example, Zheng, Chen [22] suggest that the pandemic's impact on construction productivity should be evaluated since it helps practitioners plan alternative solutions.

The literature of post-COVID-19 is limited. Using 'post-COVID-19' in the relevant journals and subjects resulted in only 344 documents (see the second search string in Appendix). This chapter suggests focusing on the post-COVID-19 period by offering new solutions, resilience strategies, and developing digital technologies, suggesting new applications and use cases, particularly from technical perspectives. **Figure 2** shows that most published papers focused on social science, business management, and health issues (see the third search string in Appendix). While digital technologies play a critical role in distance communication, monitoring, tracking, and e-management, less than 10% of the papers focused on digital development from an engineering perspective.

In line with the recovery plan and post-COVID-19 resilience strategies, the following topics are suggested to be investigated in the future:

- Risk management for continuity of tasks in construction and other relevant businesses. The need for scenario modeling and simulations as well as contingency planning is increased. Some digital tools will be helpful such as data mining and visualization, machine learning, and digital twinning. There is a wide range of risks that should be considered, such as suicide risk in construction [23], labor market during COVID-19 [24], safety measures in construction projects [19], specific sectors with tunneling and underground activities [25].

- The resilience of critical infrastructure at the national or regional levels will be one of the demanding topics.

- Remote operation, particularly in transportation or construction sites, is difficult, but it is demanding. Robotics, 3D Printing [3, 26, 27], autonomous systems, computer vision, and detection algorithms [28], monitoring and sensing technologies [29–31], the internet of things (IoT), geospatial intelligence, and analytic systems [28, 32, 33], intelligent contract administration [34], smart city technologies [35, 36] and cybersecurity tools should be further developed for different tasks. The role of Internet Of Things, drones, artificial intelligence, blockchain, and 5G technologies for pandemic management in healthcare was reviewed recently by Chamola et al. [37].

- Logistics and supply chain management needs to be redefined in the future as the core element of city resilience. The recent publications focused on the food supply chain [38], but construction and housing, real estate [39], and other sectors need to redefine or design alternatives for their supply chain management.

- Energy efficiency tools and technologies and developing alternative sources are demanding and should be investigated future. Reliable energy systems with lower costs and real-time optimization gird need to be extended further.

- Improve education and skills in the AEC sector, enabling the continuity of work using advanced or complex technologies

- Investigations on cross-sector collaborations need to be conducted. These collaborations and sharing data platforms will help practitioners learn from best and worst practices and work together to improve the supply chain process and increase resilience [1].

- Strategies should be developed to maintain the emission reduction in cities. The air quality improved in many cities due to the reduction of emissions [12, 40]. For example, Li, Li [41] reported that the PM2.5, NO_2, and SO_2 emissions decreased significantly due to the reduction of their concentrations by 31.8%, 45.1%, and 20.4% in pandemic at selected case areas of the Yangtze River Delta Region in China.

Conflict of interest

The authors declare no conflict of interest.

Appendix

The first search string used in Scopus resulted in 2,185 documents. The first string is:

(TITLE-ABS-KEY (COVID-19) AND TITLE-ABS-KEY (construction OR architecture OR city OR transportation)) AND (EXCLUDE (SUBJAREA, "MEDI") OR EXCLUDE (SUBJAREA, "BIOC") OR EXCLUDE (SUBJAREA, "IMMU") OR EXCLUDE (SUBJAREA, "AGRI") OR EXCLUDE (SUBJAREA, "PHAR") OR EXCLUDE (SUBJAREA, "NURS") OR EXCLUDE (SUBJAREA, "PSYC") OR

EXCLUDE (SUBJAREA, "NEUR") OR EXCLUDE (SUBJAREA, "HEAL") OR EXCLUDE (SUBJAREA, "CENG") OR EXCLUDE (SUBJAREA, "CHEM") OR EXCLUDE (SUBJAREA, "DENT") OR EXCLUDE (SUBJAREA, "VETE") OR EXCLUDE (SUBJAREA, "Undefined")) AND (LIMIT-TO (DOCTYPE, "ar") OR LIMIT-TO (DOCTYPE, "re")) AND (LIMIT-TO (LANGUAGE, "English")) AND (LIMIT-TO (SRCTYPE, "j"))

The second string used in Scopus resulted in 344 documents. The second string is:

TITLE-ABS-KEY (post-COVID-19) AND (EXCLUDE (SUBJAREA, "MEDI") OR EXCLUDE (SUBJAREA, "BIOC") OR EXCLUDE (SUBJAREA, "PSYC") OR EXCLUDE (SUBJAREA, "NURS") OR EXCLUDE (SUBJAREA, "AGRI") OR EXCLUDE (SUBJAREA, "NEUR") OR EXCLUDE (SUBJAREA, "IMMU") OR EXCLUDE (SUBJAREA, "HEAL") OR EXCLUDE (SUBJAREA, "DENT") OR EXCLUDE (SUBJAREA, "PHAR") OR EXCLUDE (SUBJAREA, "CENG") OR EXCLUDE (SUBJAREA, "CHEM") OR EXCLUDE (SUBJAREA, "MULT") OR EXCLUDE (SUBJAREA, "PHYS") OR EXCLUDE (SUBJAREA, "MATE") OR EXCLUDE (SUBJAREA, "VETE")) AND (LIMIT-TO (DOCTYPE, "ar") OR LIMIT-TO (DOCTYPE, "re")) AND (LIMIT-TO (LANGUAGE, "English")) AND (LIMIT-TO (SRCTYPE, "j")).

The third-string used in Scopus resulted in 162 documents. The third-string is:

(TITLE-ABS-KEY (post-COVID-19) AND TITLE-ABS-KEY (construction OR architecture OR design OR "Built Environment" OR city OR transportation))

Author details

Samad Sepasgozar[1]*, José David Bienvenido-Huertas[2], Sara Shirowzhan[1] and Sharifeh Sargolzae[3]

1 The School of Built Environment, University of New South Wales, Sydney, Australia

2 Department of Building Construction II, University of Seville, Sevilla, Spain

3 Faculty of Art and Architecture, University of Zabol, Zabol, Sistan and Baluchestan, Iran

*Address all correspondence to: sepas@unsw.edu.au

References

[1] Jallow, H., S. Renukappa, and S. Suresh, *The impact of COVID-19 outbreak on United Kingdom infrastructure sector.* Smart and Sustainable Built Environment, 2020.

[2] Sepasgozar, S.M., *Digital technology utilisation decisions for facilitating the implementation of Industry 4.0 technologies.* Construction Innovation, 2020.

[3] Tahmasebinia, F., et al., *Criteria development for sustainable construction manufacturing in Construction Industry 4.0.* Construction Innovation, 2020.

[4] Shi, A., et al., *5D BIM Applications in Quantity Surveying: Dynamo and 3D Printing Technologies*, in *Smart Cities and Construction Technologies*. 2020, IntechOpen.

[5] Shirowzhan, S., et al., *BIM compatibility and its differentiation with interoperability challenges as an innovation factor.* Automation in Construction, 2020. **112**: p. 103086.

[6] Perera, U., et al., *Application of building information modelling for fire hazard management in high-rise buildings: an investigation in Sri Lanka.* Intelligent Buildings International, 2021: p. 1-15.

[7] Shirowzhan, S. and S.M. Sepasgozar, *Spatial analysis using temporal point clouds in advanced GIS: Methods for ground elevation extraction in slant areas and building classifications.* ISPRS International Journal of Geo-Information, 2019. **8**(3): p. 120.

[8] Shirowzhan, S., J. Trinder, and P. Osmond, *New Metrics for Spatial and Temporal 3D Urban Form Sustainability Assessment Using Time Series Lidar Point Clouds and Advanced GIS Techniques*, in *Urban Design*. 2019, IntechOpen.

[9] Shirowzhan, S., W. Tan, and S.M. Sepasgozar, *Digital Twin and CyberGIS for Improving Connectivity and Measuring the Impact of Infrastructure Construction Planning in Smart Cities*. 2020, Multidisciplinary Digital Publishing Institute.

[10] Sepasgozar, S.M.E., et al., *Lean Practices Using Building Information Modeling (BIM) and Digital Twinning for Sustainable Construction.* Sustainability, 2021. **13**(1): p. 161.

[11] Sepasgozar, S.M., et al., *Additive Manufacturing Applications for Industry 4.0: A Systematic Critical Review.* Buildings, 2020. **10**(12): p. 231.

[12] Sepasgozar, S.M., et al., *Methods for monitoring construction off-road vehicle emissions: a critical review for identifying deficiencies and directions.* Environmental Science and Pollution Research, 2019. **26**(16): p. 15779-15794.

[13] Sepasgozar, S., et al., *A Systematic Content Review of Artificial Intelligence and the Internet of Things Applications in Smart Home.* Applied Sciences, 2020. **10**(9): p. 3074.

[14] Sepasgozar, S.M., et al., *Delay Causes and Emerging Digital Tools: A Novel Model of Delay Analysis, Including Integrated Project Delivery and PMBOK.* Buildings, 2019. **9**(9): p. 191.

[15] Al Amri, T. and M. Marey-Pérez, *Impact of COVID-19 on Oman's Construction Industry.* Technium Soc. Sci. J., 2020. **9**: p. 661.

[16] Bsisu, K.A.-D., *The impact of COVID-19 pandemic on Jordanian civil engineers and construction industry.* International Journal of Engineering Research and Technology, 2020. **13**(5): p. 828-830.

[17] Shibani, A., D. Hassan, and N. Shakir, *The Effects of Pandemic*

on Construction Industry in the UK. Mediterranean Journal of Social Sciences, 2020. **11**(6): p. 48-48.

[18] Pathirana, L., *Effect of COVID-19 and Strategic Response: A Review on Sri Lankan Construction Industry.* SSRG International Journal of Economics and Management Studies, 2020. **7**: p. 73-77.

[19] Amoah, C. and F. Simpeh, *Implementation challenges of COVID-19 safety measures at construction sites in South Africa.* Journal of Facilities Management, 2020.

[20] Miyakawa, D., K. Oikawa, and K. Ueda, *Firm exit during the COVID-19 pandemic: Evidence from japan.* Journal of the Japanese and International Economies, 2020. **59**: p. 101118.

[21] Alzúa, M.L. and P. Gosis, *Social and economic impact of COVID-19 and policy options in Argentina.* UNDP LAC C, 2020. **19**.

[22] Zheng, L., K. Chen, and L. Ma, *Knowledge, attitudes, and practices towards COVID-19 among construction industry practitioners in China.* Frontiers in public health, 2020. **8**: p. 981.

[23] King, T.L. and A.D. Lamontagne, *COVID-19 and suicide risk in the construction sector: preparing for a perfect storm.* Scandinavian journal of public health, 2021: p. 1403494821993707.

[24] Byrne, S., et al., *The initial labour market impact of COVID-19.* Central Bank Economic Letter, 2020. **2020**(4).

[25] Wang, Z., Z. Liu, and J. Liu, *Risk Identification and Responses of Tunnel Construction Management during the COVID-19 Pandemic.* Advances in Civil Engineering, 2020. **2020**.

[26] Tahmasebinia, F., et al., *Three-Dimensional Printing Using Recycled High-Density Polyethylene: Technological Challenges and Future Directions for Construction.* Buildings, 2018. **8**(11): p. 165.

[27] Niemeläa, M., et al. *3D Printing Architectural Freeform Elements: Challenges and Opportunities in Manufacturing for Industry 4.0*. in *Proceedings of the 36th International Symposium on Automation and Robotics in Construction (ISARC)*. 2019.

[28] Liu, C., et al., *Evaluation of classical operators and fuzzy logic algorithms for edge detection of panels at exterior cladding of buildings.* Buildings, 2019. **9**(2): p. 40.

[29] Shirowzhan, S., S. Sepasgozar, and C. Liu, *Monitoring physical progress of indoor buildings using mobile and terrestrial point clouds*, in *Construction Research Congress 2018*. 2018.

[30] Sepasgozar, S.M., C. Wang, and S. Shirowzhan. *Challenges and Opportunities for Implementation of Laser Scanners in Building Construction*. in *33rd International Symposium on Automation and Robotics in Construction (ISARC 2016)*. AUBURN, ALABAMA, USA.

[31] Yang, X., et al., *Automated PPE-Tool pair check system for construction safety using smart IoT.* Journal of Building Engineering, 2020. **32**: p. 101721.

[32] Shirowzhan, S., et al., *Data mining for recognition of spatial distribution patterns of building heights using airborne lidar data.* Advanced Engineering Informatics, 2020. **43**: p. 101033.

[33] Shirowzhan, S. and S. Lim. *Autocorrelation statistics-based algorithms for automatic ground and non-ground classification of Lidar data*. in *ISARC. Proceedings of the International Symposium on Automation and Robotics in Construction*. 2014. IAARC Publications.

[34] McNamara, A.J. and S.M. Sepasgozar, *Intelligent contract*

adoption in the construction industry: Concept development. Automation in Construction, 2021. **122**: p. 103452.

[35] Shirowzhan, S., W. Tan, and S.M. Sepasgozar, *Spatial big data, BIM and advanced GIS for smart transformation: city, infrastructure and construction*. 2020: MDPI.

[36] Zhang, K. and S. Shirowzhan, *Smart Cities and Construction Technologies*. 2020.

[37] Chamola, V., et al., *A comprehensive review of the COVID-19 pandemic and the role of IoT, drones, AI, blockchain, and 5G in managing its impact.* Ieee access, 2020. **8**: p. 90225-90265.

[38] Singh, S., et al., *Impact of COVID-19 on logistics systems and disruptions in food supply chain.* International Journal of Production Research, 2020: p. 1-16.

[39] Uchehara, I., et al., *Impacts and risk management of COVID-19 pandemic on real estate supply chain.* International journal of real estate studies, 2020. **14**(Special Issue 1).

[40] Sepasgozar, S.M. and J. Blair, *Measuring non-road diesel emissions in the construction industry: a synopsis of the literature.* International Journal of Construction Management, 2019: p. 1-16.

[41] Li, L., et al., *Air quality changes during the COVID-19 lockdown over the Yangtze River Delta Region: An insight into the impact of human activity pattern changes on air pollution variation.* Science of The Total Environment, 2020. **732**: p. 139282.

Section 2

Sustainability

Chapter 2

Rebars for Durable Concrete Construction: Points to Ponder

Anil K. Kar

Abstract

Reinforced concrete is the number one medium of construction. It is important to have good quality concrete and reinforcing bar (rebar). It is equally important to have competent bond between rebar and concrete. About six decades ago ribbed rebars of high strength steel started replacing plain round bars of mild steel, the use of which had made reinforced concrete constructions durable. It was overlooked that ribbed rebars of carbon steel would be highly susceptible to corrosion at accelerated rates. That would not only make reinforced concrete constructions reach states of distress early, that could also destroy or reduce bond between ribbed rebars and concrete. The continued use of ribbed rebars of high strength carbon steel demonstrates a widespread lack of understanding of the phenomenon of bond between rebars and concrete. This lack of understanding of bond has led to the introduction of epoxy coated ribbed rebars, ribbed stainless steel bars and glass fiber reinforced and granite reinforced polymer rebars, all of which permit reinforced concrete carry static loads because of engagement between such rebars and concrete. But the load-carrying capacity of reinforced concrete elements is impaired, and such elements become vulnerable to local or even total failure during vibratory loads. The use of PSWC-BAR, characterized by its plain surface and wave-type configuration, permits the use of medium strength and high strength steel. In the absence of ribs, the rate of corrosion is greatly reduced. The use of PSWC-BARs, at no added effort or cost, in lieu of conventional ribbed bars, leads to enhancement of effective bond or engagement between such rebars and concrete, thereby leading to increased load-carrying capacity, several-fold higher life span, ductility and energy-absorbing capacity, and great reduction in life cycle cost and adverse impact of construction on the environment and the global climate. In keeping with a lack of understanding of bond between rebars and concrete, there is arbitrariness in the selection of the required level of percent elongation and ductility of rebars.

Keywords: bond, corrosion, deformed bar, durability of concrete constructions, percent elongation, PSWC-BAR, reinforcing bar, ribbed rebar

1. Introduction

Reinforced concrete is the number one medium of construction, in which reinforcing bar (rebar) is one of the two component elements; the other element being concrete.

It was in the mid-nineteenth century when builders in different countries experimented with concrete, reinforced with steel elements of different types.

Easy availability of the component materials, easy formability, rigidity, strength, safety and durability of reinforced concrete construction made more and more people interested in such constructions.

Plain round bars of mild steel became the standard rebar.

The time-dependent performance of concrete structures, reinforced with such bars, set the standards of performance in the context of durability.

Besides the external elements, e.g., water/moisture, oxygen, carbon dioxide, chlorides, sulphates, alkalis, and other deleterious materials, which can have destabilizing effects on concrete constructions, it cannot be overlooked that the intrinsic properties of the two principal constituent materials, viz., concrete and rebars, have much to do with durability of reinforced concrete; Kar [1].

Besides concrete and rebar, "bond" between concrete and rebar, though not a material by itself, and though no one buys it or pays for it like they buy or pay for concrete and rebar, is a property that is no less important than concrete and rebar are in the context of reinforced concrete construction.

Very little consideration has been given to what leads to good "bond", and what can prevent "bond" between concrete and reinforcing elements. Also important can be the selection of an appropriate percent elongation, better still, ductility, of the material of the rebar.

In the context of "bond" and its influence on the performance of reinforced concrete, Kar [2] has suggested three terms, viz., "bond", "effective bond" and "engagement". While the last two are synonymous, that cannot be said of "bond".

Kar [2] has shown that the quality of "engagement" between rebar and concrete can greatly influence the performance of reinforced concrete elements and structures.

Buoyed by the performance of reinforced concrete, with plain round bars as rebars, engineers thought of making reinforced concrete constructions more economical by using rebars of higher strength steel.

Gradually, many different types of round reinforcing bars were introduced; Abrams [3].

Forgetful of earlier unsatisfactory experiences in the nineteenth and early twentieth century with bars, having different types of protrusions on the surface, engineers decided that the use of high strength steel would be possible by increasing the bond between rebar and concrete by providing ribs on the surface of such rebars.

Plain round bars of mild steel thus gave way to rebars of high strength steel wherein the bars are characterized by the presence of ribs on the surface (**Figures 1** and **2**). Ribbed bars were introduced in the belief that ribbed surfaces would increase bond between rebars and concrete.

The provision of ribs on the surface of rebars of high strength steel was facilitated in 1947 by ASTM International [4] publishing ASTM A305, that provided Specifications on rebar deformation patterns.

Contrary to the beliefs and expectations that (a) the presence of ribs on the surface of rebars of steel would increase the "bond" between rebars and the surrounding concrete, and (b) there would be no detrimental effect of the ribs on the performance of concrete constructions, which may be reinforced with ribbed rebars, the presence of ribs on the surface of rebars may create void spaces, at isolated locations, between rebars and concrete, thereby decreasing "bond". However, the wedge action of ribs, together with the reduced "bond", may (or may not) lead to an increase in the "engagement" between rebars and concrete.

No thought was spared as to the likely consequences the use of bars, with surface deformations or ribs, could have on the long term performance, or even on the immediate performance and load-carrying capacities of reinforced concrete constructions; Kar [1, 2, 5].

Figure 1.
Typical cold twisted deformed (CTD) rebar, with lugs and protrusions on the surface and stresses beyond yield on the entire body, which replaced plain round bars starting the decade of the 1960's.

Figure 2.
Typical high strength TMT rebars with surface deformations, which replaced plain round bars starting the decade of the 1960's.

Engineers and manufacturers of rebars blindly followed the lead of ASTM International. The Bureau of Indian Standards (BIS) in India published the Standard IS 1786 on High Strength Deformed Steel bars and Wires for Concrete Reinforcement --- Specification [6].

Though plain round bars, as in IS 432 (Part I) [7], and Grade A bars in IS 2062 [8] were available, gradually plain round bars gave way to ribbed bars where the strength of steel in rebars was increased artificially by twisting the bars beyond yield at a cold state, giving rise to CTD bars (**Figure 1**).

With time, manufacturers of rebars in India and elsewhere adopted the technique of increasing strength through the centuries-old practice of quenching, couched in diplomatic language as thermomechanical treatment, giving rise to TMT bars (**Figure 2**).

During the last sixty years or so, almost all reinforced concrete constructions worldwide have been with ribbed rebars of high strength steel, whether of the CTD or TMT type or not.

The time-dependent performances of concrete structures (**Figures 3–5**), reinforced with these later day rebars, failed to match the time-dependent performance of concrete structures, which were reinforced with plain round bars of mild steel.

The relatively poor performance of concrete structures since the introduction of high strength rebars, with surface deformations, has caused worldwide concern.

Figure 3.
Distress in staging of overhead water reservoir due to corrosion in rebars.

Figure 4.
Abandoned hospital building a decade after construction in the new township of Salt Lake City, Kolkata.

Figure 5.
Typical distress in ground level columns caused by rust in ribbed TMT bars in a 10 year old building in Kolkata.

There had to be reasons, and the reasons were not unknown; Alekseev [9, 10], and Kar [1, 5, 11–17]. But engineers and manufacturers of rebars paid no heed.

The rebars, with surface deformations, are today covered by the Indian Standard IS 1786 [6] for high strength deformed steel bars. The Standard covers both CTD and TMT bars. ASTM International in the USA published quite a few Specifications

(a) : Conventional rebars with plain surface as used to be used decades earlier; also as in PSWC-BAR

(b) : Rebars with ribs showing start of corrosion preferentially at the ribs of untwisted bars (four bottom bars) and corrosion all over the surface when the remaining length of the same bar was twisted beyond yield at a cold state (four top bars) (CTD bars)

(c) : Absolutely fresh ribbed TMT bars stacked at the factory before despatch, showing onset of corrosion all over the surface

(d) : Fresh ribbed TMT bars (IS 1786) showing start of corrosion at ribs

(e) : Corroded TMT rebars (IS 1786) in pier under construction for flyover; products of corrosion streaking down faces of pier

52-storey buildings under construction

(f) : Ribbed TMT bars coated with products of corrosion even before concreting (designer one of the top designers in the world • contractor one of India's best • rebars manufactured by one of India's best known manufacturers • there was quality control by a team of engineers from a university)

(g) : Loose rust led to greatly lowered "engagement", i.e., "effective bond", causing through-the-thickness cracks in reinforced (with ribbed TMT bars in the figure above) concrete shear walls for tall buildings in Kolkata even before a floor was completed (project has top designer, proof design consultant, leading rebar manufacturer, top contractor and quality monitoring agent) in 2011

(h) : The underside of Bridge No. 1014 up (before repairs) between Khurda Road and Palasa Stations under East Coast Railway, India

(i) : The underside of a bridge between Khurda Road and Palasa under East Coast Railway, India

Figure 6.
A collection of plain bars free from rust and ribbed CTD and TMT bars with various stages of corrosion.

on ribbed rebars of high strength steel. The most commonly used rebars are covered in ASTM A615/A615M [18].

In terms of durability, the structures may be adversely affected because of the inability of concrete to stand up to the external elements, e.g., chlorides, sulphastes, etc. or even to water as its presence may permit alkali-silica reaction in concrete in certain cases.

Most often, the durability of concrete constructions is adversely affected by corrosion in the steel rebars in the case of reinforced concrete (**Figures 3–6(h)** and **(i)**), and by corrosion in the wires and strands of steel in the case of prestressed concrete.

Though less frequent, corrosion in ribbed rebars (**Figure 2**), used as secondary reinforcement in prestressed concrete constructions, can trigger unacceptable conditions of distress in prestressed concrete constructions.

The focus here is on rebars and durability of reinforced concrete constructions, as influenced by rebars.

2. Observations following

2.1 The use of ribbed bars

Following the use of ribbed bars of high strength steel, the world has seen a significant fall in the long term performance of reinforced concrete constructions. Sights of decay and distress in concrete constructions, reinforced with ribbed rebars of steel, became inescapable (**Figures 3–5**) within years of their construction.

A 1999 survey of bridges and buildings of reinforced concrete construction in the public domain in and around Kolkata, India revealed that while none of the structures, built since the 1940s with plain round bars of mild steel, showed any sign of distress, all the structures built with ribbed bars (**Figure 1**) in the 1970s and 1980s were showing signs of distress; Kar [11].

In a 1991 article in ACI Materials Journal, American Concrete Institute, Papadakis, Vayenas and Fardis [19] wrote: "The last two decades have seen a disconcerting increase in examples of the unsatisfactory durability of concrete structures, specially reinforced concrete ones."

Sixteen years later in 2007, Swamy [20] from UK was more forthright in his expression when he wrote in the Indian Concrete Journal: "The most direct and unquestionable evidence of the last two/three decades on the service life performance of our constructions and the resulting challenge that confronts us is the alarming and unacceptable rate at which our infrastructure systems all over the world are suffering from deterioration when exposed to real environments."

An analysis of the observations by Papadakis et al. [19], by Swamy [20] and by others leads to the recognition that the relatively poor performance of reinforced concrete constructions followed the start of use of ribbed rebars of high strength steel.

Figure 5 shows typical conditions of concrete columns, reinforced with ribbed rebars (**Figure 2**), ten years after the construction of a building in Kolkata. All the columns at the ground level of the building suffered a similar fate.

The findings of the 1999 survey as well as the structures in **Figures 3–5** show clearly that compared to concrete structures, reinforced with plain round bars of mild steel, concrete structures, reinforced with ribbed bars of medium strength and high strength steel, reach states of distress much earlier.

This excessive corrosion in ribbed rebars of carbon steel suggests that the susceptibility of ribbed rebars to corrosion at accelerated rates is an intrinsic nature of ribbed rebars of carbon steel.

However, there had been hesitation by engineers in recognizing that today's ribbed bars were highly susceptible to corrosion at accelerated rates, and this excessive corrosion in today's rebars is due to.

a. the damages caused to the ribs at the time of provision of ribs on the surface

b. the damages caused to the ribs at the time of transportation and handling of rebars

c. the presence of ribs on the surface of today's rebars.

The hesitation to recognize ribs as a principal cause of excessive corrosion in rebars led not only to the continued condemnation of all new reinforced concrete constructions to early decay, distress and failure, but also to ASTM International, BIS and such other organizations publishing multiple Specifications/Standards on rebars as imagined solutions to the problem of early distress in reinforced concrete constructions, e.g., ASTM International publishing A775 [21] for epoxy coated ribbed bars, and on its failure to solve the problem, ASTM International A955/A955M for Deformed and Plain Stainless Steel Bars [22], and when that did not work, ASTM International published A1055 [23] for zinc (first coat) and epoxy (2nd coat), which too has serious limitations, as epoxy coating prevents the all-important "bond" with concrete (**Figures** 7 and **8**).

The lack of "bond" can have serious consequences: (a) cracks in structures (**Figure 6(g)**), (b) lowered load carrying capacities (Kar [2]), and (c) chunks of concrete falling (**Figure 9**) or even structures collapsing (**Figure** 7).

Like epoxy coated bars, stainless steel bars too fail to solve the problem, as ribbed bars of stainless steel too may corrode under conditions of exposure of concrete structures to chlorides, and additionally such bars may not bond or may not bond well with concrete.

Failing to recognize that the problem of early distress in today's reinforced concrete constructions is due to the use of ribbed rebars of steel as in the Indian Standard IS 1786 [6], BIS published the Indian Standard IS 13620 [24] for Fusion Bonded Epoxy Coated Reinforcing bars.

Just as BIS failed to recognize that the problem of early distress in reinforced concrete constructions started with the use of ribbed bars as in IS 1786, BIS also

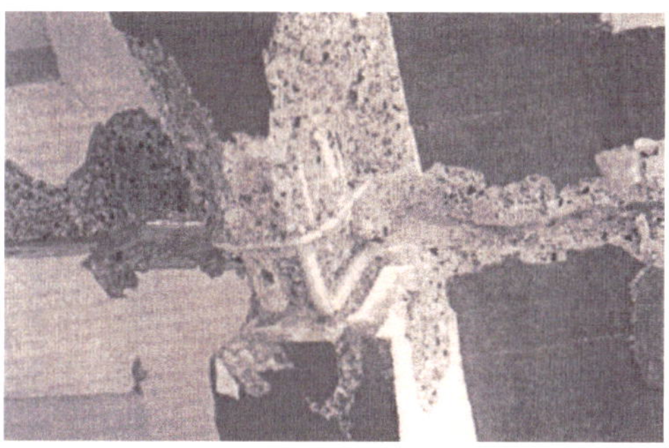

Figure 7.
Concrete easily separates from epoxy coated rebars under vibratory loading conditions whereas all structures are required to resist vibratory loads due to earthquakes; separation led to failure of buildings.

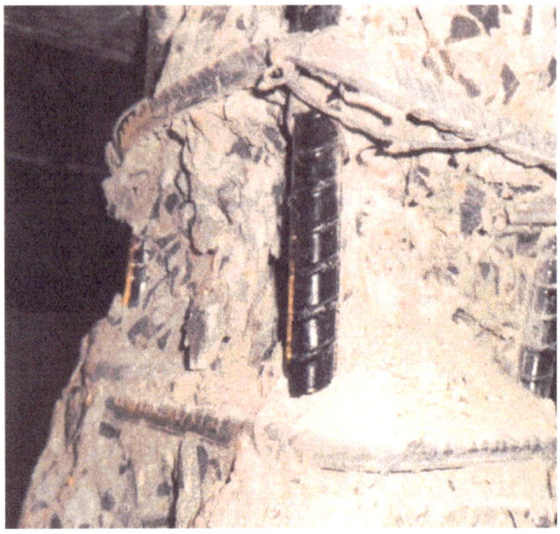

Figure 8.
The bond between epoxy coated rebar and concrete will be negligible, as seen in a column; the ribs on the surface of rebars engage the concrete up to a limit and that too when the loading is monotonous; absence of bond led to lower load-carrying capacity.

Figure 9.
A view of the deck of the Jogeswari flyover in Mumbai seven years after construction; concrete separated from rebars with poor bond qualities.

failed to recognize that, as cautioned in SubSection 5.6.1 of its Standard IS 456 [25], epoxy coated bars would not bond with the surrounding concrete, whereas the availability of the required "bond" is an essential requirement for reinforced concrete.

Similarly, as ASTM International published Specifications on epoxy coated bars and stainless steel bars, without a recognition or understanding of the basic cause(s) of early distress in reinforced concrete constructions of recent decades, and the significance of "bond" between rebars and the surrounding concrete, BIS in India followed suit by publishing the Indian Standard IS 16651:2017 on High Strength Deformed Stainless Steel bars and Wires for Concrete Reinforcement Specification [26].

The story is the same in many other countries.

It is recognized here that:

a. corrosion in rebars is greatly influenced by the intrinsic nature of the particular rebars; e.g., stainless steel bars will not generally corrode whereas mild steel and medium tensile steel bars will corrode, and high tensile strength steel bars with higher carbon contents will corrode more and at faster rates

b. the surface conditions/features on the rebar influence the rate of corrosion; the provision and the presence of ribs, as in bars conforming to IS 1786 [6] and ASTM A615/A615M [18] lead to acceleration in the rate of corrosion; Alekseev [9, 10], and Kar [1, 5, 12]

c. the manufacturing process influences the rate of corrosion; by stretching/stressing the bars beyond yield, the CTD process leads to corrosion at accelerated rates; the TMT process too hastens corrosion due to stresses from quenching effort; Alekseev [9, 10], and Kar [1, 5, 11, 12].

3. Intrinsic susceptibility of ribbed bars to corrosion

It has been recognized earlier that the problem of early distress in reinforced concrete structures started showing up following the start of use of steel reinforcing bars with ribs on the surface.

Figure 6(d) shows the start of corrosion at the ribs of TMT bars.

Figure 6(c) shows corrosion all over the surface of relatively fresh ribbed TMT bars.

The four bottom bars in **Figure 6(b)** show the start of corrosion preferentially at the ribs of untwisted ribbed bars while the four top bars show corrosion all over the surface of the ribbed bars as a consequence of stressing the bars beyond yield.

These show that:

a. the provision and presence of ribs invite corrosion

b. high stresses, specially stresses beyond yield, lead to corrosion at accelerated rates.

It cannot be overlooked that the ribs were provided out of a perceived necessity of improved "bond" between rebar and concrete when the rebars were upgraded from low-carbon to medium carbon or high carbon steel for higher strength. The truth is that the presence of ribs on the surface of rebars decreases "bond" between rebars and concrete. But the ribs may provide greater resistance to longitudinal movement of the bars relative to the surrounding concrete. Also, as found in the preceding, the ribs encourage corrosion in rebars; Alekseev [9] and Kar [1, 5].

Whether of the CTD or TMT type, or not, the reasons for ribbed bars of carbon steel being intrinsically susceptible to corrosion at accelerated rates are:

1. residual stresses develop at the bases of ribs during the manufacturing stage

2. cracks or surface damages, which trigger corrosion, may develop at the ribs at the time of manufacture, during transportation and handling

3. nominal stresses in ribbed rebars under load are enhanced in keeping with the phenomenon of stress concentration due to the presence of ribs or cracks

4. additional stresses develop in ribs in a loaded structure due to the wedge action of such ribs against surrounding concrete

5. the sum-total of stresses and strains in Items 1 to 4 approach or reach yield stress or strain levels

6. the rate of corrosion increases with increasing stress levels; the rate accelerates as the stress or strain approaches yield levels, and the surface becomes unstable once at or beyond yield, whereupon the bars become incapable of being passivated and consequently the process of corrosion becomes unstoppable; Kar [1].

The CTD and TMT processes are in violation of the inherent nature of steel to be ductile and to protect itself; Kar [1].

These CTD and TMT bars of high strength steel have another shortcoming to contend with: "The effect of stresses on corrosion is reflected more distinctly in the mechanical characteristics of the reinforcement, specially of high-strength steels with low ductility." [[10], pp. 203–204].

On the basis of extensive work in Russia, Alekseev [10] commented on the above scenario thus: "the durability of reinforcement specimens with a stepped (deformed) profile may be roughly an order less than that of smooth specimens since the former have stress concentrators on the surface at the bases of projections, which represent sites of preferential formation of cracks." [[10], pp. 221–222].

The preceding explains the reasons behind the intrinsic susceptibility of ribbed bars of steel to corrosion at accelerated rates.

It is the effect of this high susceptibility of ribbed bars to corrosion that led to the observations by Papadakis et al. [19] and Swamy [20], and to the types of early distress in reinforced concrete constructions, as depicted in **Figures 3–5**.

4. Solutions to early distress in concrete constructions

It has been recognized that rebars with surface deformations corrode excessively, leading to concrete constructions with such rebars reaching states of distress early.

The obvious solution to the problem would have been to use plain round bars as in the past. But engineers, having used in design and construction rebars of medium strength and high strength steel over the decades, would not like to go back to the use of rebars of steel having yield strength of 40 to 50 percent of the yield strength of steel in today's rebars.

Two options are available.

OPTION 1: WATERPROOFING TREATMENT.

In recognition of the fact that the problem of early distress, cited in the preceding, resulted primarily from a combination of two factors:

a. extra susceptibility (compared to that of plain round bars of mild steel) of ribbed bars, high yield strength deformed bars, and ribbed CTD and TMT bars to corrosion

b. availability of a moist environment inside concrete

and in further recognition that the problem of early distress in reinforced concrete structures could be avoided by preventing a moist environment inside concrete, Kar [13, 16, 27, 28] developed effective, practical and durable waterproofing systems and the corresponding specifications for waterproofing treatment to virtually all types of concrete structures, the implementation of which would prevent absorption of water/moisture into concrete, as in the cases of buildings, bridges and similarly exposed structures, or would prevent migration of water/moisture through the treated surface, as it would be in the cases of basements, tunnels, etc.

This waterproofing system has also the capacity to prevent the ingress of CO_2 and O_2 into the structure.

The concept of making concrete structures durable through surface protection in the nature of waterproofing treatment is gradually gaining ground in the USA and in other countries, and BIS, in recognition that concrete constructions with ribbed bars, as in IS 1786 [6], required extra protection against corrosion in the rebars, made waterproofing treatments a requirement for durability. SubSection 8.2.1 of IS 456:2000 [25] partly reads: "The life of the structure can be lengthened by providing extra cover to steel, by chamfering the corners or by using circular cross-section or by using surface coatings which prevent or reduce the ingress of water, carbon dioxide or aggressive chemicals."

It needs to be noted here that the provision of waterproofing treatments to concrete structures became essential because of the failure of the ribbed CTD and TMT bars, conforming to IS 1786 in India, ASTM A615/A615M [18] in the USA or bars conforming to similar other Standards/Specifications in other countries, to make concrete structures as durable as those used to be when the rebars had plain surfaces, and high strengths in the rebar materials were not achieved through the highly detrimental processes of cold twisting beyond yield as in the case of CTD bars (**Figure 6(b)**) or through quenching/thermal hardening/thermomechanical treatment as in the case of TMT bars (**Figure 6(c)** and **(d)**).

Kar's [16, 27, 28] art of making reinforced concrete structures durable through the provision of waterproofing treatment on the surface of such structures is an indirect way of solving the problem that was or that is invited with the use of the potentially damaging ribbed rebars of high strength steel, that was encouraged by ASTM International, BIS, ISO and such other organizations, which recommended and permitted the use of ribbed rebars, with or without the added processes of (a) cold twisting, as in CTD bars, or (b) quenching as in TMT bars, in the false belief or hope that concrete structures, reinforced with such bars, would be at least as durable as concrete structures of earlier era, which were reinforced with plain round bars of mild steel.

Though surface protection systems have worked pretty well, it does have the following shortcomings:

a. this additional treatment requires additional project time and expenditure

b. the materials used, and the specifications followed, may not be appropriate

c. there can be shortcomings in workmanship

d. such external treatments may be damaged or may have limited life spans, requiring repeat treatment

e. it does not solve the problem of excessive corrosion on the surface of rebars prior to concreting (**Figure 6(c)** and **(f)**), leading to reduction or total loss of bond between rusted rebars and concrete whereas the availability of competent "bond" between rebars and the surrounding concrete is a pre-requisite for successful performance of reinforced concrete construction.

In spite of these shortcomings, it is essential that all concrete structures, reinforced with ribbed rebars of steel, as in IS 1786 [6], ASTM A305 [4] or conforming to other Standards, be provided with surface protection in the nature of waterproofing treatment; Kar [12, 13, 16, 27, 28].

OPTION 2: PSWC-BAR AS A SOLUTION.

A better solution to the problem of early distress in reinforced concrete constructions with conventional rebars of medium strength and high strength steel would be to use plain round bars as it used to be before the 1960s or 1970s.

That would have solved the problem of excessive corrosion in rebars, and that would have made reinforced concrete constructions as durable as such constructions used to be in the past.

But the problem is that the requirement of much longer development/anchor length might not have permitted the use of plain round bars of medium strength and high strength steel.

With the innovative concept of PSWC-BAR, Kar [14] provided a direct solution (at no added effort or cost) to the problem of early distress in concrete constructions with ribbed rebars of high strength carbon steel. PSWC-BAR was initially named as C-bar.

Kar [5] explained why PSWC-BAR is the most ideal rebar for reinforced concrete constructions.

The use of PSWC-BAR, at no added effort or cost, not only solves the problem of early distress in reinforced concrete constructions through several-fold enhancement of life span of such constructions, it also enhances several fold the ductility and energy-absorbing capacity of reinforced concrete constructions; Kar [2].

The several-fold enhancement of life span, at no added effort or cost, has the effect of lowering the life cycle cost of reinforced concrete construction to a fraction of what it is today.

The use of PSWC-BAR increases load-carrying capacities of reinforced concrete elements, and through the several-fold enhancement of life span, the use of PSWC-BAR minimizes the harmful effects of construction on the environment and the global climate through considerable lowering of the need for the manufacture of cement, steel, etc. Kar [29].

PSWC-BAR, characterized by its **plain surface** and **wave-type configuration** (**Figure 10**), solves the problem of early distress in reinforced concrete constructions that can result from the use of conventional ribbed bars of medium strength and high strength steel, by eliminating initiation of corrosion at the roots of ribs.

PSWC-BAR, because of the absence of ribs or any other special surface feature, if made of the same steel, will not corrode more than conventional plain round bars would do.

PSWC-BAR, because of its gentle wave-type configuration, enhances "effective bond", i.e., "engagement" between rebar and concrete; Kar [2]. Tests on beams and columns at different universities have shown that, among all types of rebars, PSWC-BAR, with its wave-type configuration, provides the best "engagement" between rebar and concrete, leading to significant enhancement of the various positive attributes of reinforced concrete; Kar [2, 17, 30, 31] and Varu [32].

While the test for loose rust and bond, or say, loss of bond, may lead to disqualification of most or all ribbed bars, conforming to IS 1786, and such other Standards, numerous tests on beams and columns have consistently shown that among rebars of steel, the use of PSWC-BAR, free from the ill effects of ribs, and if manufactured as Grade A of Hot Rolled Medium and High Tensile Structural Steel, as in IS 2062 [8], or conforming to appropriate Standards for plain round bars, can lead to the best load-carrying capacities, ductility and energy-absorbing capacity; Kar [2], indicating thereby that the "effective bond" is the best in the case of PSWC-BARs.

Besides these big fundamental differences between today's ribbed bars, as in IS 1786, and PSWC-BARs (**Figure 10**) as in IS 2062 [8], there lies the undisputedly

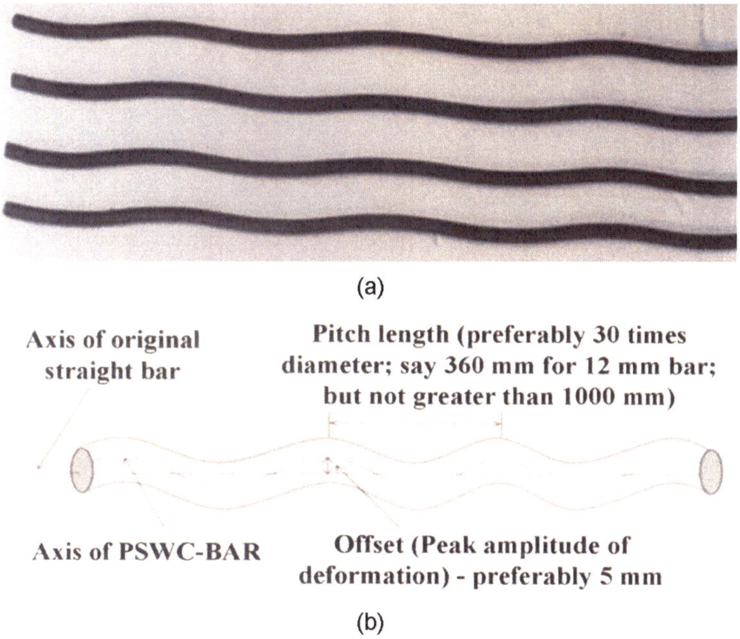

Figure 10.
PSWC-BAR of steel, characterized by plain surface and gentle wave-type configuration. (a) typical PSWC-BARs of steel, characterized by plain surface and gentle wave-type configurations. (b) schematic view of a typical PSWC-BAR.

stark difference between the very poor time-dependent performances (durability) of concrete structures, reinforced with ribbed bars, as in IS 1786 [6], ASTM A615/A615M [18] and such other Standards/Specifications elsewhere and the time-dependent performances of concrete structures, reinforced with hot rolled plain round bars with wave-type configuration, which are characteristic of PSWC-BARs.

There are various other advantages of using PSWC-BAR as rebars in reinforced concrete construction. A comparison of the load–displacement plots in **Figure 11(a)** and **(b)** show clearly that:

a. because of several fold higher ductility and energy-absorbing capacity, the use of PSWC-BARs as rebars has the potential to prevent structural failures and catastrophes during earthquakes

b. because of several times higher deflection (displacement) of flexural elements, there can be visible warnings before failure, thereby saving lives.

c. load-carrying capacities of reinforced concrete elements increase when PSWC-BARs are used.

Recommended mechanical properties of PSWC-BAR for durable and earthquake resistant concrete constructions are provided in **Table 1**.

Kar [5, 14–17] has written extensively on PSWC-BAR, and, encouraged by the many benefits, which the use of PSWC-BARs can provide, students at different universities have written a number of theses on the relative performances of concrete elements, reinforced with PSWC-BARs and conventional rebars.

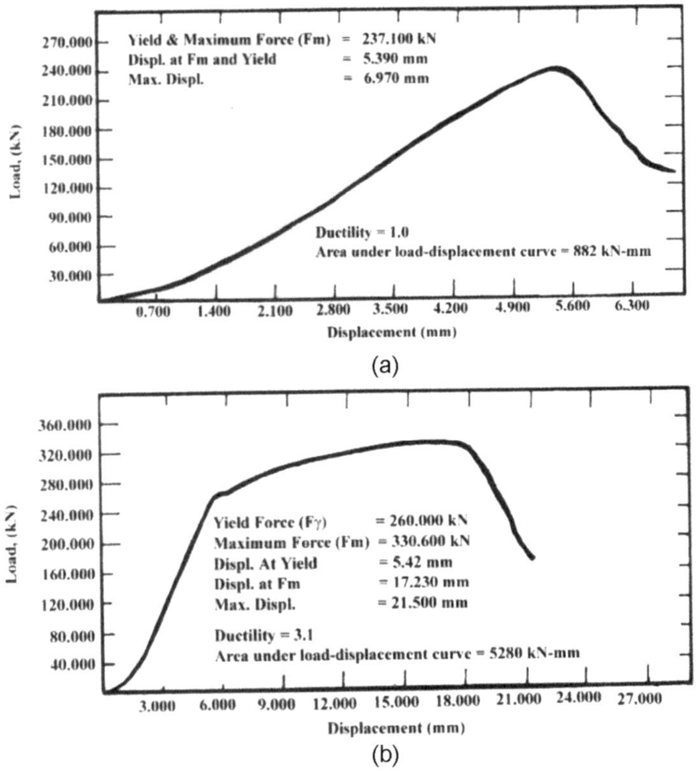

Figure 11.
Ductile response of beam reinforced with PSWC –BAR. (a) Load–displacement plot of beam with conventional rebars showing failure as the stress in rebars reached the yield stress level. (b) Load–displacement plot of beam with PSWC-BARS showing failure as the stress in rebars went past yield and approached the ultimate. Note: The two plots in (a) and (b) are drawn to different scales.

Sl. No. (1)	Property (2)	Fe 415 (3)	Fe 500 (4)	Fe 550 (5)
i)	yield stress, *Min*, N/mm^2	415.0	500.0	550.0
ii)	yield stress, *Max*, N/mm^2	500.0	600.0	660.0
iii)	$Y/Y_{specified}$ ratio[1]	1.02–1.2	1.02–1.2	1.02–1.2
iv)	$TS/Y_{specified}$ ratio[2]	≥ 1.15 - ≤ 1.40	≥ 1.15 - ≤ 1.40	≥ 1.15 - ≤ 1.40
v)	Elongation, percent, *Min*. on gauge length $5.65\sqrt{A}$, where A is the cross-sectional area of the test piece	20.0	16.0	12.0

Note: [1)] $Y/Y_{specified}$ ratio refers to ratio of actual yield strength to specified yield stress of the test piece.
[2)] $TS/Y_{specified}$ ratio refers to ratio of tensile strength to specified yield stress of the test piece.
Additional Note: [1)] The steel shall be suitable for welding processes.

Table 1.
Mechanical properties of steel in PSWC-BARs.

5. Bond in reinforced concrete

Bond between rebars and their surrounding concrete is of utmost importance in the context of reinforced concrete.

This bond, when adequately developed, permits composite response of reinforced concrete through effective transfer of forces between concrete and rebar.

Any reduction in bond, below a certain level, will lead to a reduction, or in extreme cases, a total loss of load-carrying capacities of the constructed structures, as it happened during the Bhuj earthquake on 26 January 2001 when three buildings, reinforced with epoxy coated bars, collapsed 300 kilometers away in Ahmedabad, India (**Figure 7**).

In the case of plain rebars of mild steel or carbon steel, when free from the damaging effects of the ribs as well as the CTD and TMT processes, there will be chemical bond between the mortar in concrete and the hard adherent products of very limited corrosion on the steel material, as in the cases of plain round bars of mild steel or, better still, PSWC-BARs, conforming to plain round rods of Grade A of structural steel in the Indian Standard IS 2062 [8], in which case the rods are given the wave-type configuration (**Figure 10**) at the end of the rolling mill process; Kar [14].

Similarly, PSWC-BARs can be made to conform to provisions in existing Standards/Specifications for plain round bars in other countries. Alternatively, Standards may be specifically prepared for PSWC-BARs.

The chemical bond between the mortar in concrete and the hard and adherent products of corrosion on the surface of PSWC-BARs develops shear capacity at the interface of concrete and the rebar for the transfer of forces, through shear, from concrete to rebars.

In the context of reinforced concrete, this is the "bond" engineers have been familiar with.

This should suggest that, technically speaking, there can be no "bond" between concrete and a painted surface, like the surface of an epoxy coated bar (**Figure 8**), or similarly between concrete and a stainless steel bar.

The same situation can develop if there will be loose rust on the surface of rebars as in the case of ribbed CTD or TMT bars (**Figure 6(f)**), as in IS 1786, which are the most widely used rebars in India.

Figure 6(g) shows that the loss of bond rendered the reinforcement, that was provided for load-carrying requirements, insufficient even as minor temperature reinforcements, and thereby led to the development of through-the-thickness shrinkage cracks in the shear walls even though it was a well-engineered project, except that, as per conventional practices in India, ribbed bars, as in IS 1786 [6], are used without the required scrutiny for "bond", that is set in SubSection 5.6.1 of IS 456 [25].

This is what happened in the case of the ribbed TMT bars in **Figure 6(f)** even when the bars were manufactured by a leading manufacturer of rebars and other products of steel in India.

There is more to "bond".

It is recognized that manufacturers/sellers of epoxy coated and stainless steel bars may not agree to the suggestion that there is no "bond" between epoxy coated or stainless steel bars and the surrounding concrete.

In the absence of any reliable test method to measure "bond" or bond strength in the cases of ribbed bars, engineers too tend to agree with manufacturers and sellers of epoxy coated and stainless steel bars, and they might even suggest that their tests have shown that the bond strength of epoxy coated bars is sixty percent or even eighty percent of that of uncoated bars.

The observations by engineers may be right, but their claims on "bond" are wrong. There are various reasons for it.

There is generally no "bond" between concrete and epoxy coated or stainless steel bars (**Figures 7–9**).

Any resistance to pull-out forces in the case of epoxy coated ribbed bars or ribbed stainless steel bars is essentially due to the wedge action of ribs embedded in concrete.

In the present context of bond, the epoxy coating on fusion bonded epoxy coated bars, as in IS 13620 [24], ASTM A775 [21], ASTM A934/A934M [33], ASTM A1055 [23] and similar Standards/Specifications on epoxy coated bars in other countries can be thought of as "coats of paints" as noted in SubSection 5.6.1 of IS 456 [25].

Recognizing that coats of paints, like loose rust, oil, etc. could destroy or at least reduce "bond", IS 456, the basic reinforced concrete code of practice in India, has put words of caution in SubSection 5.6.1 of its Section **5.6 Reinforcement** thus: 5.6.1 "All reinforcement shall be free from loose mill scales, loose rust and coats of paints, oil, mud or any other substances which may destroy or reduce bond. Sand blasting or other treatment is recommended to clean reinforcement."

In construction with fusion bonded epoxy coated rebars in India or elsewhere, no sand blasting or other treatment is provided so as to meet the requirements set in IS 456 or in any other document, and so as to ensure that there would be competent and adequate bond between such bars and the surrounding concrete.

It is possible that in recognition of this reality, IS 456 in its Section **5.6 Reinforcement** did not consider epoxy coated bars, as in IS 13620 [24], or stainless steel bars, as in IS 16651 [26], for possible use as rebars in reinforced concrete construction.

Though IS 456, the basic Indian Standard for reinforced concrete construction, does not approve of the use of epoxy coated bars as in IS 13620 [24] and stainless steel bars as in IS 16651 [26], such bars, which do not bond with concrete, with attended shortcomings in the performance of concrete constructions, do find use in reinforced concrete constructions in India and elsewhere.

In a series of tests by Varu [32] on thirtythree reinforced concrete columns at Nirma University in Ahmedabad, India, nine columns were reinforced with epoxy coated bars; of which three columns were with epoxy coated plain round bars, three columns were with epoxy-coated ribbed TMT bars of the type in IS 1786 [6], and three columns were with epoxy coated PSWC-BARs.

There is no suggestion that PSWC-BARs and conventional plain round bars may ever be given epoxy coating for protection. But in the test program these bars too were given epoxy coating just to have a more comprehensive understanding of the influence of surface coating (see SubSection 5.6.1 of IS 456 [25]) on load-carrying capacities and "bond" or "engagement".

The full details will be found in the thesis by Varu [32]. The observations can also be found in a few articles; Kar [2], and Kar, Dave and Varu [30].

Among other observations, it was observed:

a. unlike in the cases of the twentyfour columns with uncoated rebars of different types, there were clear indications at the failure region of all the nine columns with epoxy coated rebars that there was no bond of concrete/concrete mortar with the epoxy coated bars. A typical case is seen in **Figure 8**.

b. the epoxy coated bars led to failure of the columns at loads which were less than the loads at which the other similarly constructed, but with uncoated bars of same/similar manufacture had failed. It appeared that the coated bars did not participate in sharing loads on the columns; Kar et al. [30].

In the absence of any bond, the use of epoxy coated and stainless steel bars will lead to under-performance of reinforced concrete elements; Kar et al. [30] and Kar [2], and the use of such bars can lead to unacceptable consequences during vibratory loads (**Figure 10**), specially during earthquake events (**Figure 8**), as it happened when several multi-storey buildings in Ahmedabad collapsed on 26 January 2001 during the earthquake at Bhuj 300 km away.

The failures occurred due to separation between epoxy coated rebars and the surrounding concrete (**Figure 7**).

These should be proof enough that any claim of 60–80 percent "bond" between epoxy coated bars and concrete is wrong.

This should suggest that all concrete structures which were constructed with fusion bonded epoxy coated rebars, are suspect. In other words,

a. the margin of safety in structures with epoxy coated ribbed bars is less than what it may be thought to be as per conventional design; Kar [2] suggested modification to current design practices by considering the "effective bond" or "engagement" instead of assuming that there is competent "bond" between epoxy coated rebars and concrete.

b. all concrete structures, reinforced with epoxy coated bars, remain specially vulnerable against vibratory loads, including earthquakes, as evidenced in the failure of structures in Ahmedabad during the Bhuj earthquake of 26 January 2001.

In the cases of rebars, with ribs on the surface, where a certain amount of resistance to slippage is available, it is partly due to "bond" and partly due to the interlocking of the ribs with the surrounding concrete. From an engineering point of view, this resistance to slippage may preferably be referred to as "effective bond" or "engagement", instead of "bond".

Thus, though in the context of reinforced concrete, engineers have traditionally used only one term, i.e., "bond", and though in the context of reinforced concrete, where the rebar is a conventional plain bar of mild steel or carbon steel (**Figure 6(a)**), the use of the term "bond" may not create any confusion, the terms "effective bond" and "engagement" may be the more appropriate terms in the case of ribbed bars (**Figures 2** and **6(b)** and **(c)**) and PSWC-BARs (**Figure 10**), ribbed stainless steel bars, ribbed epoxy coated bars, polymer coated glass fiber reinforced bars, etc.

In the case of a PSWC-BAR, devoid of ribs or any other surface feature, there will be the "bond" on the entire surface, and in addition, the wave pattern along the length of the bar will provide physical resistance to slippage. The sum total of the "bond" and the "physical resistance" in the case of a PSWC-BAR can be termed as "effective bond" or "engagement".

Tests on numerous reinforced concrete beams and columns, with reinforcing bars of different types, at different universities have consistently shown that "the effective bond" or "engagement" is the highest in the case of PSWC-BARs, leading to the highest load-carrying capacities as well as several hundred percent higher ductilities and energy-absorbing capacities compared to the cases of conventional bars without the wave-type configuration; Kar [2].

In the context of reinforced concrete, there should thus be a recognition of "effective bond" or "engagement", and a clear understanding of "bond".

For similar reasons, the use of the term "engagement" will hopefully avoid a false belief that there is bond between stainless steel bars and the surrounding concrete, and it will hopefully avoid the type of collapses of reinforced concrete bridges and buildings that Ahmedabad was witness to during the earthquake of 26 January 2001, 300 kilometers away at Bhuj (**Figure 7**).

There are instances where chunks of concrete fell down from bridge decks which were constructed with ribbed TMT bars as in IS 1786 [6]. **Figure 9** shows one such example.

It should help put a stop to the use of not only the conventional epoxy coated bars, as in IS 13620 [24], but also to bars where the top coat is with epoxy as in

ASTM A1055 [23], and also to stainless steel bars as in ASTM A955/A955M [22] and IS 16651 [26], as, unlike in the cases of low carbon steel bars, stainless steel bars will not develop a thin layer of strong adherent rust on their surface for bonding with mortar in concrete.

Also, these bars stand in the way of composite response of concrete and the embedded bars, because of which even the capacity to carry static loads would be less than those which would have been arrived at on the basis of conventional design practices; Kar et al. [30] and Kar [2].

In the context of bond, besides the information provided hereinabove, Kar [14] had suggested that in the case of ribbed bars, coarse aggregates could in places rest on/against neighboring ribs (**Figure 12**), thereby blocking mortar from bonding with rebars, and also preventing passivation of rebars at such isolated locations. The void spaces aid the cause of corrosion.

In their tests, Mohammed, et al. [34] too observed void spaces beneath ribbed bars, resulting in higher rates of corrosion in ribbed bars than in the case of plain bars.

Whether in India or abroad, it has been the practice to assume that the use of ribbed bars provides the required bond between such bars and the surrounding concrete.

Though the presence of ribs on the surface of bars decreases the "bond", when compared to the cases of plain bars, the presence of ribs on the surface of bars may in some cases increase the "engagement".

Figure 6(g) presents a case where the absence of "bond" led to a decrease in the "engagement" between rebar and the surrounding concrete.

To start with, ribs were provided on the surface of rebars of high strength steel with an intent to increase bond between such rebars and concrete. This act boomeranged as it led to an acceleration in the rate of decay in reinforced concrete constructions.

The high strength in steel was/is gained in some cases either through the twisting of the bars beyond yield at a cold state or through quenching. The provision and the presence of the ribs, coupled with the twisting beyond yield or the quenching, lead to corrosion at unacceptably accelerated rates on the surface of the rebars; Alekseev [9, 10], and Kar [1, 5, 11–17] (**Figure 6(b)** and **(f)**), resulting in reduction or total destruction of the "bond" (**Figure 6(g)**). While the immediate effect of the

Figure 12.
Barrier effect of ribs, lugs and protrusions on the surface of ribbed rebars of steel preventing cement mortar from coming in contact with rebar.

destruction of "bond" is visible in **Figure 6(g)**, the long term effects are visible in **Figures 3–6(h)** and **(i)**.

Besides questionable "bond", the ribbed CTD and TMT bars, as in IS 1786 [6], meant for use as rebars in reinforced concrete construction, may not be permitted to be used as rebars, as because, such bars, with high susceptibility to corrosion at accelerated rates, will in many or most cases, fail the qualification test for rebars which have been set in SubSection 5.6.1 for reinforcement in IS 456 [25].

An example will be found in **Figure 6(g)** where it is seen that in the construction of six 48–52 storeyed buildings at a site, the shear walls, which in the absence of columns, were reasonably reinforced, developed through-the-thickness shrinkage cracks, about a metre apart as excessive loose rust on ribbed TMT bars (**Figure 6(f)**), prevented/destroyed "bond" between concrete and the highly rusted fresh rebars.

Visits to construction sites revealed that easily visible through-the-thickness shrinkage cracks in new constructions were very common. This lack of "bond" can lower the load-carrying capacities of such constructions.

The bars, conforming to IS 1786, were thus unfit for construction, at least in the light of the requirements in SubSection 5.6.1 of IS 456.

In the face of all the problems of insufficient "bond" in the case of ribbed rebars of high strength steel, epoxy coated ribbed bars, ribbed bars of stainless steel, and unacceptably high rate of corrosion in rebars, conforming to IS 1786, PSWC-BAR of medium tensile and high tensile steel (**Table 1**), conforming to IS 2062 [8], or to any other appropriate Standard/Specification for plain round bars of carbon steel of high strength steel, stands out as the only bar of high strength steel that is free from the varied problems of all other bars of high strength steel.

PSWC-BAR, endowed with the property of best "engagement", i. e., "effective bond" with concrete, also stands out as the only bar, the use of which, besides several-fold enhancement of life span, increases, by several hundred percent ductility and energy absorbing capacity of reinforced concrete construction (**Figure 11**) and Kar [2].

It is apparent that there has not been a clear understanding of the phenomenon of "bond" between rebar and concrete, what creates this "bond", what can affect the development of "bond", and what are its roles in the performance of reinforced concrete.

It is because of this lack of understanding of "bond" and its significance that made manufacturers and sellers of rebars, designers of reinforced concrete structures, construction engineers, and officials of BIS and such other organizations, who put the stamp of approval on ribbed rebars, overlook all these years the reality, the cautions in text books and Standards which read something like: all reinforcement shall be free from loose mill scales, loose rust and coats of paints, oil, mud or any other substance which may destroy or reduce bond.

It is this total failure to recognize the many significances of "bond" in the realm of reinforced concrete that facilitated the unchecked use of ribbed bars in reinforced concrete construction all these years, and in the process caused very significant losses to property owners, and great harm to the national wealth of countries, as well as to the environment and the global climate.

The facts, that (a) ribbed bars, conforming to IS 1786 and to Standards/ Specifications on ribbed bars in other countries, are highly prone to the development of loose rust on the surface of such rebars, (**Figure 6(f)**), (b) this rust can "destroy or reduce bond" between concrete and rebars (**Figure 6(g)**), (c) without competent bond between rebar and concrete there cannot be reinforced concrete in its true sense, and (d) the loose rust will prevent any possible passivation of rebars by the alkaline pore water in concrete, and thus stand in the way of protection of rebars against corrosion unless concrete constructions will be given surface protection in the nature of waterproofing treatment, have not sunk into the minds of all those who should have known, are obvious from the continued poor performance of the structures in **Figures 3–6(h)** and **(i)**, and uncounted other structures which have been and are being constructed with ribbed bars.

Kar [2] has shown that besides success and failure, and besides the issue of durability, the "effective bond" or "engagement" between rebars and the surrounding concrete may influence the load-carrying capacity, ductility and energy-absorbing capacity of reinforced concrete elements.

6. Percent elongation of rebar

Percent elongation is an important measure of ductility of rebars, that can influence the performance of the rebar and in turn the performance of concrete elements under load as well as under exposure to the environment; Kar [14]. The percent elongation is of course a very important property that may greatly influence the survivality of reinforced concrete constructions during earthquake events.

In recognition of the fact that the changing material compositions and manufacturing processes, as well as the increasing yield strengths of rebar materials during recent decades, are generally associated with decreasing percent elongation, the Specifications of ASTM International and the Standards of BIS allow/permit the use of rebars with smaller percent elongation properties with increasing yield strength of the rebar material.

It is recognized here that there are certain differences between the gage/gauge lengths in the ASTM and BIS test specimens. However, these differences do not substantially affect the following observations on percent elongation.

ASTM A615/A615M [18] of 12 Jan, 2016 has set the minimum percent elongation of rebars for Grades 75, 80 and 100, i.e., yield strengths of 520 MPa, 550 MPa and 690 MPa, to 7 percent for rebars having diameters up to 25 mm, and an even lower 6 percent for rebars having diameters greater than 25 mm, whereas for Grade 40 (280 MPa) and Grade 60 (420 MPa) bars, ASTM sets the minimum percent elongation at 12 and 9, respectively.

Similarly, IS 1786 [6], through its Amendment No. 03, dated 19-09-2017, has set the minimum percent elongation at 10.0, 10.0 and 10.0 for rebars of yield strengths 600 MPa, 650 MPa and 700 MPa, whereas it has set allowable percent elongations at 14.5 to 18.0 for different varieties of 415 MPa bars, and 12.0 to 16.0 for different types of 500 MPa bars.

Several questions arise, viz.,

a. if once it is recognized that the percent elongation of the steel material for rebars is an important and thus an inviolable property, that is to be set for acceptability of rebars, then why smaller percent elongation properties (as 6 in ASTM A615/A615M [18] and 10 in IS 1786 [6]) be considered permissible for higher yield strength materials, but not for smaller yield strength materials?

b. or, are the percent elongation properties, set in the Specifications/Standards violable, and the set properties merely represent values which certain manufacturers can achieve in the cases of bars they make?

c. how is it that when the achievable (with reasonable effort) percent elongation gets smaller and smaller with increasing yield strength, ASTM A615/A615M

[18] has set the same elongation at 7 percent or 6 percent for steel having yield strengths of 520 MPa, 550 MPa and 690 MPa?

d. If 6 percent elongation is considered acceptable for 690 MPa steel, then why should such a low percent elongation be not acceptable in the cases of rebars with steel of lower yield strengths?

e. how is it that, when the achievable (with reasonable effort) percent elongation gets smaller and smaller with increasing yield strength, IS 1786 [6] has set the same figure of 10 percent for rebars having yield strengths of 600 MPa, 650 MPa and 700 MPa?

f. how is it that when ASTM A615/A615 M [18] finds it difficult to achieve percent elongation greater than 6 for 600 MPa hot rolled bars, IS 1786 finds a 10 percent elongation achievable for 700 MPa TMT bars, when it is known that, compared to hot rolled processes, as in the USA, the TMT process, as in India, leads to hardening and lowering of ductility and percent elongation properties?

There needs to be a clear understanding of the significance of percent elongation and or ductility of rebars in the context of performance of reinforced concrete elements.

It may be desirable to set, irrespective of the yield strength of steel, a single value, below which the percent elongation or ductility will not be acceptable in the cases of rebars of steel.

In view of the fact that virtually all structures in India and in many other countries are required to be earthquake resistant, a reasonably high value may have to be set for the required percent elongation or ductility of rebars.

In this conflicting scenario, with a view to minimizing the rate of corrosion and also to improve ductility and energy absorbing capacity, PSWC-BAR, conforming to IS 2062, and possessing the property of improving "effective bond" over and above the normally available "bond", with a minimum percent elongation of 16, is recommended as the rebar of choice. The yield stress will be limited to a maximum of 550 MPa, preferably to 500 MPa; Kar [5].

Greater details on the development and mechanical properties of PSWC-BAR, together with design aid, so as to take advantage of the power of PSWC-BAR to enhance load-carrying capacity, as well as ductility and energy-absorbing capacities of reinforced concrete elements, are provided in the article: The Search for an Ideal Rebar for Durable Concrete Construction Leads to PSWC-BAR; Kar [5].

7. Ductility ratio

A better measure of the mechanical property of a rebar, and that of the performance of a concrete flexural element, reinforced with such a bar, would have been the ductility ratio rather than the arbitrarily selected percent elongation.

Assuming that the percent elongation will be at least large enough to ensure that the specified yield strength and the specified ultimate strength of the bar will be achieved, the only other useful information that a percent elongation may provide is a vague understanding that the bar may not break during necessary bending.

That should suggest that vaguely specified percent elongation is an unnecessary specification when separate tests for bending of bars are specified.

In contrast, while tests for yield and ultimate strengths (stresses) will ensure the said strengths (stresses), the information on ductility and the shape of the load-deformation plot of the bar beyond yield will provide important information on an idea about the bendability of a rebar. And in addition, the ductility ratio, coupled with a plot of the load-elongation curve of the bar will provide a great deal of information about the performance of a flexural element beyond the yield stress level of the rebar, provided that the rebar will have the requisite "engagement" with concrete, and it happens best in the case of PSWC-BARs; Kar [2].

8. Summary

Reinforced concrete is the number one medium of construction. Besides strength, easy formability and availability of the constituent materials, trouble-free long term performance, i.e., durability of concrete structures, constructed with plain round bars of mild steel, having yield stress of around 250 MPa to 280 MPa, had helped reinforced concrete attain this position.

It has been suggested that, in the context of reinforced concrete, besides concrete and rebars, "bond" between such rebars and the surrounding concrete deserves equal consideration.

Engineering practice shows that though there is a need for a clear understanding of "bond", and though the ensurement of adequate "bond" is an essential necessity, these are almost totally neglected.

Similarly, the important property of percent elongation or ductility of the rebar has not been considered with the thoroughness it deserves.

With time, besides significant changes in properties of cement, a constituent component of concrete, the reinforcing bar (rebar) was gradually changed from plain round bars of mild steel to plain round bars of medium tensile steel (yield stress of about 350 MPa) and then on to today's ribbed rebars of high strength (yield stress 415 MPa to about 700 MPa) steel.

The use of ribbed rebars of high strength steel, susceptible to corrosion at accelerated rates, led to concrete structures reaching states of distress early.

In consideration of durability, ribbed bars, as in IS 1786 in India, and ASTM A615/A615M in the USA and as in such other Standards/Specifications elsewhere, should thus be avoided.

The high susceptibility of ribbed rebars to corrosion may in cases destroy or reduce "bond" between concrete and ribbed rebars of high strength steel.

Such bars may not stand scrutiny for eligibility for use as rebars for reinforced concrete construction. It has been shown that PSWC-BAR, characterized by its plain surface and wave-type configuration, is the most ideal rebar for reinforced concrete construction.

While the plain surface of PSWC-BARs would ensure that the susceptibility of such bars to corrosion will be several orders of magnitude less than the susceptibility of conventional ribbed bars of high strength steel, the wave-type configuration of PSWC-BARs ensures that the "bond" or "engagement" between such bars and the surrounding concrete is no less than the "bond" between ribbed rebars and concrete.

Numerous tests on concrete beams and columns, reinforced with PSWC-BARs, and with ribbed bars, conforming to IS 1786, have consistently revealed that the "effective bond" or "engagement" between PSWC-BARs and the surrounding concrete is greater than the "effective bond" between concrete and ribbed rebars, conforming to IS 1786.

It is this greater "effective bond" that increases the load-carrying capacity, ductility and energy absorbing capacity of concrete elements, reinforced with PSWC-BARs.

The use of PSWC-BAR, characterized by its plain surface and wave-type configuration, at no added effort or cost, can solve the worldwide problem of early distress in reinforced concrete construction.

Besides several-fold enhancement of life span, with many added benefits, like greatly reduced life cycle cost, the use of PSWC-BAR increases by several hundred percent the ductility and energy-absorbing capacity of flexural elements of reinforced concrete. It may thus prevent catastrophes during earthquakes.

Recommended mechanical properties of PSWC-BARs for durable concrete constructions are provided.

In consideration of requirements for durability and resistance to earthquake forces, the yield stress of steel in PSWC-BAR is recommended to be limited to 550 MPa, and preferably to 500 MPa.

The several-fold enhancement of life span of concrete structures, with the use of PSWC-BARs, instead of conventional ribbed bars, can prevent staggering financial losses to property owners and to national economies of all countries as well as great harm to the environment and to the global climate.

An alternative way to enhance the durability of reinforced concrete construction is to provide, at additional cost, surface protection in the form of waterproofing treatment to concrete structures.

Author details

Anil K. Kar
Engineering Services International, Salt Lake City, Kolkata, India

*Address all correspondence to: dr.anil.kar@gmail.com

References

[1] Kar, A. K., "Concrete Structures — the pH Potential of Cement and Deformed Reinforcing Bars", paper; *Journal of The Institution of Engineers (India)*, Civil Engineering Division, Volume 82, Kolkata, June 2001, pp. 1-13.

[2] Kar, A. K., "A Theory on the Performance of Reinforced Concrete Elements," Proceedings of the Institution of Civil Engineers — Journal of Construction Materials, https://doi.org/10.1680/jcoma.18.00019, pp. 1-9.

[3] Abrams, D. A., "Tests of Bond between Concrete and Steel", Bulletin No. 71, published by the University of Illinois, Urbana, Vol. XI, No. 15, December 8, 1913.

[4] ASTM Standard A305, Specification for Minimum Requirements for the Deformations of Deformed Steel Bars for Concrete Reinforcement, ASTM International, 1947.

[5] Kar, A. K., "The Search for an Ideal Rebar for Durable Concrete Construction Leads to PSWC-Bar," International Journal of Technical Innovation in Modern Engineering & Science (IJTIMES); Volume. 5, Issue 09, September 2019, pp. 76-89.

[6] Indian Standard 1786:2008 - High Strength Deformed Steel Bars and Wires for Concrete Reinforcement --- Specification, Amendment No. 3 (Fourth Revision), Bureau of Indian Standards.

[7] Indian Standard IS 432 (Part 1):1982 (Third Revision), Specification for Mild Steel and Medium Tensile Steel Bars and Hard-Drawn Steel wire for Concrete Reinforcement, Bureau of Indian Standards.

[8] Indian Standard IS 2062:2011, Hot Rolled Low, Medium and High Tensile Structural Steel, Bureau of Indian Standards.

[9] Alekseev, S.N., "Corrosion of Steel Reinforcement," Durability of Reinforced Concrete in Aggressive Media, Oxford & IBH Publishing Co. Pvt. Ltd., (Translation of: Dolgovechnosti zhelezobetona v aggressivnikh sredakh, Stroiizdat, Moscow), (1990), Chapter 7, pp. 164-247.

[10] Alekseev, S. N., "Corrosion of Steel Reinforcement in Concrete," Chapter 10 in Moskvin, V. (edited by), translated from the original by V. Kolykhmatov, "Concrete and Reinforced Concrete Deterioration and Protection," 1990, English translation, Oxford & IBH Publishing Co. Pvt. Ltd. New Delhi, 1993, original Mir Publishers, Moscow 1990, pp. 242-260.

[11] Kar, A. K., "Deformed Reinforcing Bars and Early Distress in Concrete Structures", Highway Research Bulletin, Number 65; Highway Research Board, Indian Roads Congress, New Delhi; December 2001, pp. 103-113.

[12] Kar, A. K., "Woe Betide Today's Concrete Structures", New Building Materials & Construction World; New Delhi; Vol. 13, Issue-8, February 2008, pp. 132-158.

[13] Kar, A. K., and Vij, S. K., "Enhancing the Life Span of Concrete Bridges", New Building Materials & Construction World, New Delhi, Vol. 15, Issue 6, December 2009, pp. 134-156.

[14] Kar, A. K., "Improved Rebar for Durable Concrete Constructions", New Building Materials & Construction World; New Delhi; Vol. 16, Issue-1, July 2010, pp. 180-199.

[15] Kar, A. K., "Rebar for Durable Bridge and Other Concrete Constructions",

Indian Highways; Vol. 39, No. 3; The Indian Roads Congress (IRC), New Delhi; March 2011, pp. 59-65.

[16] Kar, A. K., "Durability of Concrete Bridges and Viaducts", The Masterbuilder; Vol. 14, No. 7, Chennai; July 2012, pp. 110-130.

[17] Kar, A. K., "A Reinforcing Bar for Durable Concrete Constructions and Much More," The Masterbuilder; Vol. 20, No. 9, Chennai; September 2018, pp. 136-146.

[18] ASTM Standard A615/A615M, Standard Specification for Deformed and Plain Carbon-Steel Bars for Concrete Reinforcement, ASTM International, 1968 (original), 2015.

[19] Papadakis, V.G., Vayenas, C.G., and Fardis, M. N., "Physical and Chemical Characteristics Affecting the Durability of Concrete", ACI Materials Journal, American Concrete Institute, March–April, 1991.

[20] Swamy, R. N., "Infrastructure Regeneration: the Challenge of Climate Change and Sustainability — Design for Strength or Durability", The Indian Concrete Journal, The ACC Ltd., Vol. 81, No. 7, July , (2007) 7-13.

[21] ASTM Standard A775, Specification for Epoxy-Coated Steel Reinforcing Bars, ASTM International, 1992.

[22] ASTM Standard A955/A955M, Specification for Deformed and Plain Stainless Steel Bars for Concrete Reinforcement, ASTM International, 1996.

[23] ASTM Standard A1055, Standard Specification for Zinc and Epoxy Dual-Coated Steel Reinforcing Bars, ASTM International, 2008.

[24] Indian Standard IS 13620:1993 (Reaffirmed 2004), Fusion Bonded Epoxy Coated Reinforcing Bars --- Specification (Fourth Revision), Bureau of Indian Standards.

[25] Indian Standard IS 456:2000, Plain and Reinforced Concrete --- Code of Practice (Fourth Revision), Bureau of Indian Standards.

[26] Indian Standard 16651:2017, High Strength Deformed Stainless Steel Bars and Wires for Concrete Reinforcement Specification (First Revision), Bureau of Indian Standards.

[27] Kar, A. K., "Making Concrete Bridges Durable Through Surface Protection by Waterproofing", The Masterbuilder; Vol. 18, No. 6, Chennai; June 2016, pp. 76-82.

[28] Kar, A. K., "Surface Protection Waterproofing of Segmental Concrete Bridges and Viaducts," The Masterbuilder; Vol. 21, No. 6, Chennai; June 2019, pp. 126-130.

[29] Kar, A. K., "Mitigation of GHG Emission and Assault on the Climate and the Environment with the use of PSWC-Bar – an Ideal Rebar for Durable Concrete Construction," The Masterbuilder; Vol. 21, No. 9, Chennai; September 2019, pp. 102-110.

[30] Kar, A. K., Dave, U. V., and Varu R. S. "Performances of columns reinforced with PSWC-bars and other rebars," *The Indian Concrete Journal*, July 2018, Vol. 92, No. 7, pp. 12-17.

[31] Kar, A. K., Dave, U. V., and Patel N. A., "Performances of Beams Reinforced with PSWC-Bars and Other Rebars," Journal of Structural Engineering, CSIR-Structural Engineering Research Centre, Chennai, Vol. 46, No. 1, April–May 2019, pp. 29-38.

[32] Varu, R. S., "Studies on C-bar as Reinforcement for Columns", a thesis submitted to Department of Civil Engineering, Institute of Technology,

Nirma University, Ahmedabad, May 2014.

[33] ASTM Standard A934/A934M, Specification for Epoxy-Coated Prefabricated Steel Reinforcing Bars, ASTM International, 1995.

[34] Mohammed, T.U., Otssuki, N., and Hisada, M., "Corrosion of Steel Bars with Respect to Orientation", ACI Materials Journal, American Concrete Institute, March–April 1999, pp. 154-159.

Chapter 3

Climate Resilience, Megalopolis Vulnerability and Spatial Distribution

Tao Ma, Nairong Tan, Xiaolei Wang, Hao Wang and Mingxi Zhou

Abstract

This chapter takes three megalopolises including Guangdong-Hong Kong-Macao, Yangtze River Delta and Beijing-Tianjin-Hebei as research objects, firstly analyzes the connection relationship and megalopolis vulnerability among core cities in the context of regional integration. Secondly, we calculate there megalopolises to obtain the vulnerability of each city in 2018 Sex index. The results show that the central cities and economically underdeveloped cities of the three megalopolises are relatively vulnerable areas in the urban agglomerations, and areas have low sensitivity and high response. Finally, policy suggestions for megalopolis are given to improve the adaptive capacity of tackling climate change. The innovation of this chapter is to use spatial data to comprehensively evaluate and analyze the vulnerability, and to realize visualization in the map, which better reflects the distribution law and proposes a response to megalopolis vulnerability.

Keywords: climate change, megalopolis, vulnerability, connectivity

1. Introduction (The identification of megalopolis vulnerability to climate change)

1.1 The concepts of vulnerability to climate change

Climate change mainly includes temperature, precipitation, solar radiation, wind speed, humidity, and air pressure. It represents a typical multi-scale global change problem, characterized by the infinite diversity, multiple pressures and time scales [1]. There is much research on vulnerability caused by climate change, and also has gradually become a forward position field of global environmental change and sustainable development. Vulnerability includes the fields of disasters research (delineated into human ecology, hazards, and the 'Pressure and Release' model) [2], which is the core of climate change. It can include the vulnerability of natural systems, such as the impact of floods and hurricanes caused by climate change on agricultural production and human life; it also includes economic and social vulnerability [3], emphasizing risks and uncertainties come from economic and social environment change. In social vulnerability, it is often measured as a function of the socioeconomic conditions of the communities [4].

1.2 The identification of megalopolis vulnerability

A city is a complex giant system, and urban systems have been growing exponentially in size and complexity [5]. While a megalopolis is a complex coupled giant system. Spatial scales about connection relationships and measurement indicators in quite different ways. The energy use of industry, infrastructure (water, electricity, gas), land and buildings in cities are the importance of factors in climate change [6]. Megalopolis is core to cities, which the highly integrated connectivity networks formed by developed infrastructure networks such as transportation, information and communications. With the industrial division, factor flow, and the integration of cross-regional infrastructure in megalopolis, there are more connectedness in cities, while the more vulnerability increased due to the effects of climate change.

The concept of urban vulnerability evolved from vulnerability and refers to the ability affecting by adverse events. It was first proposed by the United Nations Development Program in 1999, and specifically includes sensitivity and capacity of response [7]. The reasons for urban vulnerability include the natural environment and external shocks related to trade and diplomacy. At present, the research on urban vulnerability in China mainly focuses on the resource-based cities (such as mining cities and ocean cities).

There are more than 660 cities in China, including 7 megalopolises with urban populations of more than 10 million, 9 megacities with populations of 5-10 million, and 124 large cities with populations of 1 to 5 million. The five major urban agglomerations (the Pearl River Delta, the Yangtze River Delta, the Beijing-Tianjin-Hebei, the middle reaches of the Yangtze River, and Chengdu-Chongqing) gather 55% of the country's total economic output and 40% of the total population. In 2018, the domestic product of the Yangtze Delta megalopolis reached about 2.7 trillion US dollars, and the economic scale surpassed that of Italy, which is the eighth largest economy in the world, and the Pearl River Delta megalopolis was about 1.2 trillion US dollars, which is equivalent to Mexico. However, in general, the internal links are not connected between China's megalopolises and cities, it should consolidate cooperation in different sizes of cities. The urban commuter railway mileage in central cities such as Beijing and Shanghai is less than other international metropolises. Core cities, surrounding areas, and peripheral cities were of inferior spatial connectivity, and even some core cities have "fault zones" with surrounding areas.

China's megalopolises are the spatial carriers that form economies of scale and scope, they are also a crucial support for improving economic efficiency, resilience, and resistance to external uncertain risks. In particular, as the construction of new infrastructure will enter a stage of rapid growth, which represented by 5G technology, artificial intelligence and data centers. It will be beneficial to the sustainability of megalopolises and the adaptability to vulnerabilities and risks, for enhancing the connectivity, the flexible layout adjustment of cities's internal structure.

This chapter firstly identifies the vulnerabilities of Guangdong-Hong Kong-Macao, Yangtze River Delta and Beijing-Tianjin-Hebei, and establishes the sensitivity index and capacity of response index for the three megalopolises under the effects of climate change. Secondly, it assesses the impact of urban vulnerability, and finally put forward response the measures for China's megalopolises that are climate-adaptive.

2. Analysis of the vulnerability of Chinese urban agglomerations and urban continuous belts

2.1 Theoretical model

According to the concept of urban vulnerability, relevant indicators are selected to establish an urban vulnerability evaluation index system, as shown in **Table 1**.

Target layer	Criterion layer	Code	Indicator layer	Positive and negative	Weight
Urban vulnerability index(V)	Sensitivity index(S)	x_1	Industrial dust emissions	Positive	0.0857
		x_2	Industrial sulfur dioxide emissions	Positive	0.0577
		x_3	Industrial wastewater emissions	Positive	0.0575
		x_4	Fiscal deficit	Positive	0.0097
		x_5	Urban registered unemployment rate	Positive	0.0325
		x_6	Financial institution loan balance	Positive	0.1631
	Coping Ability Index(R)	x_7	Urban green coverage rate	Negative	0.0070
		x_8	Per capita disposable income of urban residents	Negative	0.0423
		x_9	Actual use of foreign capital	Negative	0.1217
		x_{10}	Freight volume	Negative	0.0754
		x_{11}	Passenger volume	Negative	0.1162
		x_{12}	Urban road area	Negative	0.0836
		x_{13}	Domestic and foreign currency household deposits of financial institutions	Negative	0.0904
		x_{14}	The proportion of science and education expenditure in fiscal expenditure	Negative	0.0572

Table 1.
Urban vulnerability evaluation index system.

Urban vulnerability refers to the sensitivity it exhibits when faced with the influence of multiple factors inside and outside the urban system and the strength of its coping ability to adapt to this sensitivity. The higher the sensitivity of the urban system, the stronger the urban vulnerability. The higher the response capacity of the urban system, the smaller the urban vulnerability. Therefore, urban vulnerability is a function of the sensitivity and coping ability of the urban system when faced with internal and external factors [8].

$$V = f(S,R) = \frac{S}{R} \tag{1}$$

In the formula, V represents the urban vulnerability index. S represents the urban sensitivity index and R represents the urban response capability index. Urban vulnerability is directly proportional to sensitivity and inversely proportional to coping ability.

According to the calculation method of the urban vulnerability evaluation index system [7], the acquired data is processed first. Due to the diversity of data types in the evaluation index system, the dimensions and magnitudes of different index data have certain differences [9]. Therefore, it is necessary to further non-dimensionalize the acquired data to eliminate its impact on urban vulnerability evaluation indicators. In this evaluation indicator system, all indicators have a positive correlation with the sub-target level [10]. The standardization method of deviation of the data is selected to process the original data without dimension. The standardization of the dispersion of the data is a linear transformation of the original data, so that the result falls into the interval [0,1]. The conversion function is as follows:

$$X_j = \frac{x_j - \min\{x_j\}}{\max\{x_j\} - \min\{x_j\}} \tag{2}$$

Second, the entropy method is used to determine the weight of each index in the urban vulnerability evaluation index system. The specific process is as follows:

1. Quantify all indicators with the same measurement. Calculate the proportion of the j-th index value of the i-th evaluation object, the calculation method is as follows:

$$p_{ij} = \frac{x_{ij}}{\sum_{i=1}^{m} x_{ij}} \tag{3}$$

2. Calculate the information entropy e_j:

$$e_j = -\frac{1}{\ln m}\sum_{i=1}^{m}\left(p_{ij}\ln p_{ij}\right) \tag{4}$$

   In the formula, $0 \le e_j \le 1$; when $p_{ij} = 0$, $e_j = 0$.

3. Calculate the difference coefficient g_i:

$$g_i = 1 - e_j \tag{5}$$

   The smaller the entropy value, the greater the difference between indicators.

4. Calculate the index weight w_j:

$$w_j = \frac{g_j}{\sum_{i=1}^{n} g_j} \tag{6}$$

2.2 The indicator system

According to the concept of urban vulnerability [11], relevant indicators are selected to establish an urban vulnerability evaluation index system, as shown in **Table 1**.

2.3 Research scope and data sources

The research scope includes the Beijing-Tianjin-Hebei urban agglomeration, the Yangtze River Delta urban agglomeration and the Guangdong-Hong Kong-Macao

City group	Cities
Beijing-Tianjin-Hebei City Group	Beijing; Tianjin; Baoding, Tangshan, Langfang, Shijiazhuang, Qinhuangdao, Zhangjiakou, Chengde, Cangzhou, Hengshui, Xingtai, Handan, Dingzhou, Xinji in Hebei Province; Anyang in Henan Province
Yangtze River Delta City Group	Shanghai; Nanjing, Wuxi, Changzhou, Suzhou, Nantong, Yancheng, Yangzhou, Zhenjiang, Taizhou in Jiangsu Province; Hangzhou, Ningbo, Jiaxing, Huzhou, Shaoxing, Jinhua, Zhoushan and Taizhou in Zhejiang Province; Hefei, Bengbu, Wuhu in Anhui Province, Ma'anshan, Tongling, Anqing, Chuzhou, Chizhou, Xuancheng
Guangdong-Hong Kong-Macao Greater Bay Area	Hong Kong; Macau; Guangzhou, Shenzhen, Zhuhai, Foshan, Zhongshan, Dongguan, Zhaoqing, Jiangmen, Huizhou in Guangdong Province

Table 2.
Definition of the study area.

Greater Bay Area. The relevant indicator data comes from the 2019 data China City Statistical Yearbook, the 2019 provincial and municipal statistical yearbooks and statistical bulletins. Due to the availability of data, the calculation does not include Hong Kong, Macau Special Administrative Region, Dingzhou and Xinji county-level cities. For individual missing values, replace with the mean value of the city group where the city is located (**Table 2**).

2.4 Result analysis

Combining the relevant index data of the studied cities and using the urban system vulnerability evaluation index model, the 2018 Beijing-Tianjin-Hebei urban agglomeration, the Yangtze River Delta urban agglomeration, and the Guangdong-Hong Kong-Macao Greater Bay Area are calculated to obtain the vulnerability index of each city in 2018. As shown in **Figure 1**.

The central cities and economically underdeveloped cities of the three major urban agglomerations are relatively vulnerable areas in the urban agglomerations, and are low-sensitive and high-response areas. The vulnerability of the urban system is divided into 4 levels according to the clustering results. Among the Guangdong-Hong Kong-Macao Greater Bay Area, Guangzhou is a very vulnerable area with a vulnerability index of 0.45, a sensitivity index of 0.1, and a coping capacity index of 0.34; Shenzhen is a low-vulnerability area; other areas are not vulnerable. Among the Yangtze River Delta urban agglomerations, Shanghai and Suzhou are vulnerable areas with a vulnerability index of 0.57 and 0.43, of which Shanghai's vulnerability index and sensitivity index are 0.18 and 0.39, respectively. Hangzhou and Nanjing are vulnerable areas. Other areas are not fragile areas. In the Beijing-Tianjin-Hebei urban agglomeration, Beijing is a very vulnerable area, with the vulnerability index and sensitivity index being 0.18 and 0.44, respectively. Tianjin and Tangshan are low-vulnerability areas. Other areas are not vulnerable areas. The vulnerability structure of the central cities of China's three major urban agglomerations is obvious. Their economic development is in a leading position in the urban agglomerations and the country, and from the perspective of urban infrastructure, cities have a strong ability to deal with vulnerability. Therefore, these cities are vulnerable. Therefore, urban sensitivity can be reduced through environmental protection policies (**Table 3**) [12].

3. Measures to improve the climate adaptability of megacities

According to the results of the second section, the fragility structure of the central cities of the three major urban agglomerations is obvious. Their economic

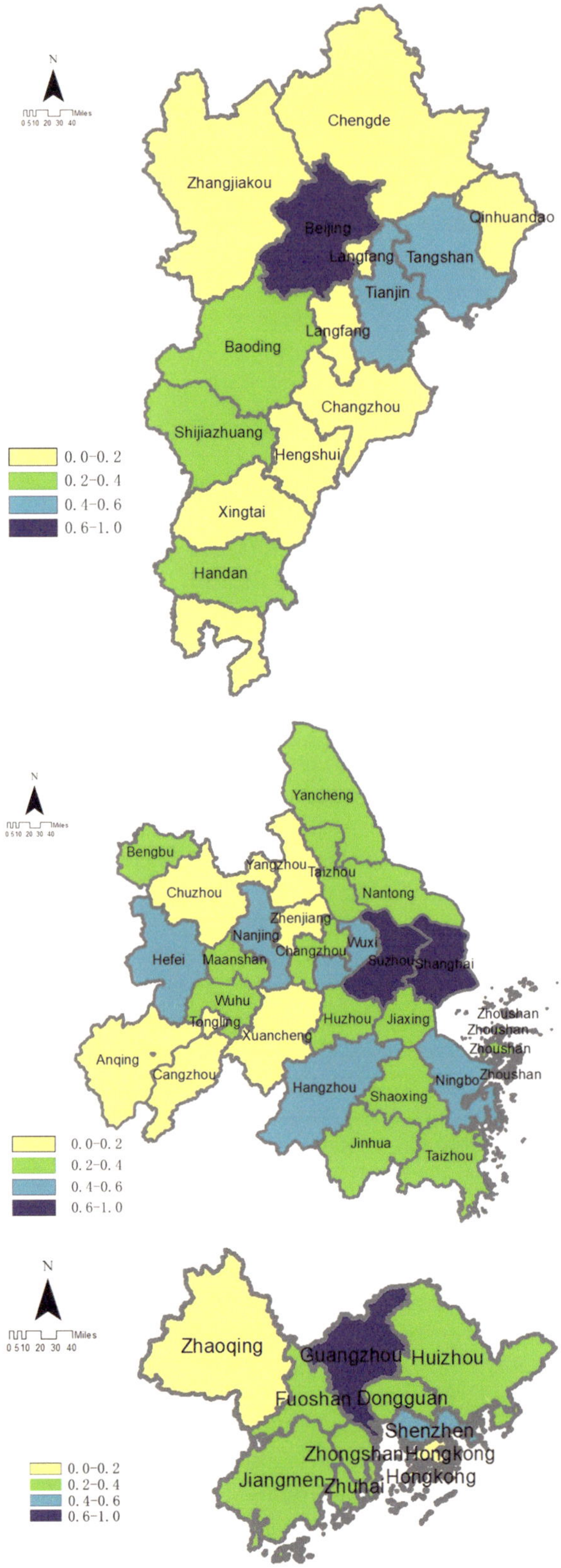

Figure 1.
Sketch map of vulnerability of Beijing-Tianjin-Hebei urban agglomeration, Yangtze River Delta urban agglomeration and Guangdong-Hong Kong-Macao Greater Bay Area.

Vulnerability value	$0 \leq V < 0.2$	$0.2 \leq V < 0.4$	$0.4 \leq V < 0.6$	$0.6 \leq V < 1$
Classification	Not vulnerable	Low vulnerable	Vulnerable	Very vulnerable

Table 3.
Criteria for the classification of urban vulnerability levels.

development is in a leading position in the urban agglomerations and the whole country, and from the perspective of urban infrastructure, cities have a strong ability to deal with fragility. The central cities and economically underdeveloped cities of the three major urban agglomerations are all relatively vulnerable areas in the urban agglomerations, and are areas with low sensitivity and high response capacity. For such cities, the focus is to improve their comprehensive carrying capacity. In addition, a series of pan-smart city technologies that emerged under the background of the fourth industrial revolution are playing an increasingly important role in the development and operation of cities. Scientifically and rationally cognizing the relationship between technology and urban multi-agents, and guiding the coordinated development of future technological development and urban planning, design, construction, and governance is crucial for building a resilient system of climate-adaptive large urban agglomerations.

3.1 Improve the comprehensive carrying capacity of central cities and urban agglomerations

Optimize the layout of population functions to enhance the city's carrying capacity. Urban agglomerations must form a reasonable allocation of population and industrial layout within the city, and form a "city-region" organizational model in terms of modern economic cultivation. Optimizing the population distribution within urban agglomerations can not only alleviate the "urban disease" caused by the excessive concentration of population in megacities, but it is also a necessary means to realize the high-level population function of world-class urban agglomerations and enhance the overall participation of the urban agglomeration in global competition. To optimize the layout of population functions with a global perspective, it is necessary to consider the differentiated needs of the diverse population structure of world-class urban agglomerations for urban space and urban functions, and improve the comprehensive carrying capacity of the urban agglomeration.

Give full play to comparative advantages and build a multi-level power system for regional development. To improve the carrying capacity of central cities, it is necessary to strengthen regional integration and build a multi-level dynamic system for regional development. According to the differences in the location conditions of the economy, society and resource environment, division of labor, coordination, and organic configuration form an overall process of synergy. The idea of planning and formulating the main functional zone can be followed. Some areas do the division of ecological products, and some areas do the division of industrial products. The main function is different, the area type will be different. Different regions have different development levels and advantages, so the focus of division of labor is also different. By improving regional policies and spatial layout, mechanisms for counterpart cooperation, cross-regional cooperative governance, transfer payments, and ecological compensation should be established between advantageous regions and underdeveloped regions.

Improve the efficiency of agglomeration and tilt the allocation of construction land to advantageous areas. In order to improve the efficiency of agglomeration and give full play to comparative advantages, it is necessary to strengthen the flexibility of construction land management. For areas with serious population loss, land use indicators can be appropriately reduced, and land use efficiency can be improved on existing land, so as to avoid idle land waste. For those areas with long-term population growth and rapid growth, and areas with tight land use, urban space can be developed by increasing land use indicators. Or expand the scope of cities through the integrated development of urban agglomerations or metropolitan areas, and use the scale of urban agglomerations or metropolitan areas to expand market space. Peripheral cities in metropolitan areas have the greatest potential and focus for enhancing their carrying capacity. In the future, apart from adjusting administrative divisions, they can also integrate the economy of metropolitan areas or urban agglomerations for integrated development, which can also promote the development of peripheral cities in metropolitan areas.

Promote green development and build a livable environment for ecological city clusters. Ecological environment is the key element and bottom line of comprehensive carrying capacity. How to achieve a win-win situation for ecology and economy, first, adjust the industrial structure and energy structure, develop green industries and use new energy, reduce energy consumption, and reduce environmental pollution. Strengthen the protection of the common ecological space, strengthen the coordinated treatment of environmental pollution such as water, gas, and soil, comprehensively treat the ecosystem of large rivers and lakes, and improve the ecological benefits and environmental carrying capacity of central cities and urban agglomerations. Determine the development goals and scale of the city based on the carrying capacity of resources and environment, delineate ecological red lines, avoid the development of urban pie, and reduce the number of healthy urban garden cities. Improve the ecological and environmental protection legal system, and improve the ecological compensation mechanism and environmental compensation mechanism.

3.2 Improve urban resilience with pan-smart city technology

Use pan-smart city technology to obtain public data. Public data acquisition is the first step in rapid and intelligent decision-making and coordinated prevention and control of urban vulnerability emergencies. Information channels and data acquisition sources mainly include urban IoT systems, communication networks and online information. When urban vulnerability events occur, the use of big data technology can realize the identification and supervision of risk areas and specific groups of people, fully clarify the flow of various types of people, and provide a large amount of data support and early warning and decision-making for the coordinated prevention and control of emergencies; In the communication network, the biological signs and infrared data recorded by the RFID sensor, the user's comprehensive life, work, entertainment and usage habits information stored by the wireless smart phone, and the voice, text, image, and video information recorded by the base station communication system are all Precise prevention and control and the coordinated prevention and control of various functional departments provide effective support; online public opinion information and Internet browsing records on the Internet can effectively reflect the media and people's public opinion, and can eliminate rumors and guide through timely and effective data information management and guidance. Avoid panic, carry out active information dissemination at the same time, and strengthen coordination in all aspects.

Use pan-smart city technology for data interaction and intelligent decision-making. There are many problems and difficulties in urban resource management, such as the personal user privacy and security needs of ordinary people, the security of business management confidential data of enterprises and institutions, and the security of confidential information of national government agencies. The distributed storage technology and consensus mechanism of the blockchain can provide reasonable and effective solutions to the above problems, and "upload" public decisions such as management and emergency to the blockchain, and arrange corresponding credit incentives, honor incentives, and material incentives In accordance with the constraints of laws and regulations, the use of artificial intelligence technology for in-depth mining of big data. On the basis of blockchain-based collaborative prevention and control, we propose a public chain for ordinary people, alliance chains for national government agencies and enterprises and institutions, and blockchain network scenarios for private chains for confidential systems. At the same time, the national policy guidelines Under the supervision of relevant supervisory agencies, formulate effective incentive mechanisms and relevant laws and regulations to promote data interaction and intelligent decision-making capabilities, and realize the efficient, safe, intelligent, reliable and authoritative of large-scale accident prevention and control of big data Reliable processing level.

Build an intelligent platform around the lifeline of the city. The urban emergency information platform is the neural network and nerve center of the emergency management system. It plays an important role in cross-departmental emergency data management, information sharing, business collaboration and comprehensive services. It is to improve the information, scientific, intelligent and integrated service of urban public crisis management. An important means of synergy. The urban emergency information platform is an open and complex giant system. Its construction is a complex social project involving many factors and problems. It is necessary to use cloud computing technology to further improve the basic structure and functions of the platform itself, and it also needs to learn from the information resource system., Emergency data standards, data technology applications, and talent team construction are fully prepared for the construction of emergency information platforms.

4. Conclusions

Combining the relevant index data of the research cities, and using the urban system vulnerability evaluation index model, we calculate the 2018 Beijing-Tianjin-Hebei urban agglomeration, the Yangtze River Delta urban agglomeration and the Guangdong-Hong Kong-Macao Greater Bay Area to obtain the vulnerability of each city in 2018 Sex index. The results show that the central cities and economically underdeveloped cities of the three major urban agglomerations are relatively vulnerable areas in the urban agglomerations, and are areas with low sensitivity and high response. According to the clustering results, the vulnerability of the urban system can be divided into four levels. In the Guangdong-Hong Kong-Macao Greater Bay Area, Guangzhou is a very vulnerable area, and Shenzhen is a vulnerable area. Among the city clusters in the Yangtze River Delta, Shanghai and Suzhou are vulnerable areas; Hangzhou and Nanjing are vulnerable areas. In the Beijing-Tianjin-Hebei urban agglomeration, Beijing is a very vulnerable area, and Tianjin and Tangshan are low-vulnerability areas. The vulnerability structure of the central cities of China's three major urban agglomerations is obvious. Their economic development is in a leading position in urban agglomerations and countries. From the perspective of urban infrastructure, cities have a strong ability to deal with vulnerabilities. To this end, this paper proposes measures to respond to climate-adaptive large urban agglomerations in terms of enhancing the carrying capacity of central cities and strengthening the application of pan-smart city technologies.

5. Discussion

This chapter is to use spatial data to comprehensively evaluate and analyze the vulnerability of major megalopolise, and to realize visualization in the map, which better reflects the distribution law of urban vulnerability, and proposes a response to urban vulnerability. In this chapter, the three megalopolise vulnerability show that the more developed the urban economy, the higher the vulnerability index. This is an interesting and very thought-provoking discovery. We think this research is valuable because we have seen the urban vulnerability in the most economically developed regions of China, where the core cities are connected to other cities by elements, industries and infrastructure. And the results show that the more developed the city is, the higher the vulnerability will be. Therefore, in the medium and long term development of China in the future, how to overcome the vulnerability and continuously enhance the resilience of megalopolise should be further put forward countermeasures. In fact, as infrastructure integration and city-to-city connectivity grow, so does vulnerability. From the perspectives of government, industrial layout, elements flow, wage and welfare, etc., it is also a question that needs further thinking about how to deal with the increasing risks in the future megalopolise in the form of layout, industrial (employment) structure and welfare growth.

Due to the limitation of data access, some descriptive indicators of urban vulnerability have not been introduced into the evaluation system, which limits the research conclusions. In the follow-up, multi-source heterogeneous data should be collected to strive for more accurate and comprehensive evaluation results.

The follow-up research needs to use regional climate simulation or statistical downscaling methods to predict regional climate change, and comprehensive simulation results provide suggestions for each city group.

Author details

Tao Ma*, Nairong Tan, Xiaolei Wang, Hao Wang and Mingxi Zhou
Harbin Institute of Technology, Harbin, China

*Address all correspondence to: matao@hit.edu.en

References

[1] Adger W N. Vulnerability, Global Environmental Change. 2006,16 (3), 268-281. DOI:10.1016/j.gloenvcha.2006.02.006

[2] B. L. Turner, Roger E. Kasperson, Pamela A. Matson, et al. A Framework for Vulnerability Analysis in Sustainability Science. 2003, 100(14):8074-8079. DOI:1073/pnas.1231335100.

[3] Susan L. Cutter,Bryan J. Boruff,W. Lynn Shirley. Social Vulnerability to Environmental Hazards. Social Science Quarterly,2003,84(2). DOI:10.1111/1540-6237.8402002.

[4] Karakoc D B , Almoghathawi Y , Barker K , et al. Community resilience-driven restoration model for interdependent infrastructure networks[J]. International Journal of Disaster Risk Reduction, 2019, 38:101228. DOI:10.1016/j.ijdrr.2019.101228

[5] André Botequilha-Leitão a, Emilio R. Díaz-Varela b. Performance Based Planning of complex urban social-ecological systems: The quest for sustainability through the promotion of resilience. Sustainable Cities and Society, 56. DOI:org/10.1016/j.scs.2020.102089

[6] Patricia R L , Daniel G , Olga W , et al. Urban Sustainability and Resilience: From Theory to Practice. Sustainability, 2016, 8(12):1224. DOI:10.3390/su8121224

[7] Nelson K , Gillespie-Marthaler L , Baroud H , et al. An integrated and dynamic framework for assessing sustainable resilience in complex adaptive systems. Sustainable & Resilient Infrastructure, 2019:1-19. DOI: 10.1080/23789689.2019.1578165

[8] LIU S Y. Research on economic vulnerability of specialized Tourist City. In: Yunnan University. 2019

[9] Vliet M T H V , Wiberg D , Leduc S , et al. Power-generation system vulnerability and adaptation to changes in climate and waterresources. Nature Climate Change, 2020:11-09. DOI: 10.1038/nclimate2903

[10] YANG Junyan, SHI Beixiang, SHI Yi, et al. Construction of a Multi-Scale Spatial Epidemic Prevention System in High-Density Cities. City Planning Review, 2020, 44(3):17-24.

[11] CHEN Weike, YAN Chaohua, DONG Jing, et al. Temporal and Spatial Dynamic Evolution and Key Fragility Factor Analysis of Urban Vulnerability: Taking Henan Province for Example. Urban Problems, 2020(3): 38-46.

[12] ZHANG Fan. The Prevention and Control of Infectious Diseases Should Be Included in the City's Comprehensive Disaster Prevention and Mitigation Plan as Soon as Possible: A Written Conversation on the 2020 Novel Coronavirus Pneumonia Emergency. City Planning Review, 2020-02-12[2020-06-07]. http://kns.cnki.net/kcms/detail/11.2378.TU.20200211.1757.008.html.

Chapter 4

Organizational Analysis of Sustainable Building Certifications in Mexico City

Sara Velasco-Baca and Fermín Cruz-Muñoz

Abstract

In 2008, the Sustainable Building Certificate Program was implemented in order to identify and to foster sustainable buildings. The aim of the work is to analyze, since an organizational perspective, the contribution of the users' "day a day" actions to archive sustainability. The principal variable is the building property actions to obtain the certification and the criteria defined at the Certification Program. The central significance of this work is to define how important is the implementation of strategies to introduce sustainable patterns to the building users in contrast with the facility features. The most interesting finding shows that certification design and punctuation assignation to each criteria influence in sustainable actions. The incorporation of technology devices primes over the sustainable actions by the users of the buildings. This condition represents a practical implication because the sustainable actions pattern presents a challenge to the sustainable vision. People do not assume the sustainability as a change in human actions, but a technological question. So, the central value of this research is to demonstrate the very low importance in the Mexico City certification program and stakeholders in sustainable user's patterns. The principal limitation the research is many sustainable buildings, certificated by international programs were not includes. This consideration implies future studies, to identify a general tendency of sustainable actions related with the users' activity.

Keywords: sustainable building, organizational analysis, sustainable certification, sustainable actions, Mexico City

1. Introduction

Buildings represent the most prevalent form of artificial use of a city's territory, and most of its natural and energy resources are used by human constructions. Therefore, buildings are a major source of environmental impact. Metropolises have been often shown to be responsible for substantial environmental impact; they completely transform the natural ecosystem into an essentially anthropic space. The creation of cities initiates processes that change soil conditions, wildlife environments, and water bodies, contributing to climate change and environmental degradation as a result of the loss of biodiversity and resources [1].

In Mexico, buildings account for 25% of total electricity consumption, 17% of energy consumption, 5% of water consumption, 20% of carbon dioxide emissions, and

20% of waste generation [2]. Most of the pollutants generated by buildings are due to the use of low-quality materials and unsustainable products, which result in constant economic costs related to the restoration and replacement of obsolete materials that degrade or become damaged. On the other hand, insufficient building maintenance results in decreased resilience to environmental problems due to the deterioration of energy and water supply and distribution equipment over time [3].

The role of habitability is essential in building design and construction processes as a leading consideration concerning the different needs of users in terms of sustainability; sustainable buildings are resilient and respond to the current environmental problems affecting habitability [4]. Consequently, many institutions created sustainability certifications in order to decrease the environmental impact of the buildings. These certifications consider a set of criteria linked with the general qualities of the buildings, most of them in terms of technology and design. Nevertheless, there is no studies about the role of the everyday user's activity linked with the sustainability as consequence of the internal organization in the buildings. Therefore, there are some criteria related with the human activity, we believe that these factors are so few in comparison with the technological and design factors. To promote sustainability patterns in user emphasis, the construction of a new consciousness and avoid delegating the sustainability as a technological question.

The purpose of this chapter is to analyze the design of the Mexico City's sustainable building certification program from an organizational perspective and to analyze its implications in actions carried out by building owners to obtain such certification in the period from 2011 to 2018. The Sustainable Buildings Certification Program seeks to promote environmental actions aimed at improving the habitability and sustainable configuration of affordable buildings [5]. The present study is focused on facilities operated by the private sector. We will analyze actions carried out by proprietary companies and users that directly determine the way in which building users organize toward sustainability. This research contributes to understand how the internal organization of users foster sustainable patterns, as a vital factor to create a sustainable consciousness in urban people.

The first analysis unit is the environmental criteria used to evaluate sustainable buildings, which will be analyzed on the basis of scores assigned to each criteria in terms of efficiency and excellence. The second analysis unit is the actions carried out by agents who adapt buildings in search of certification. We used these analysis units were to determine whether everyday sustainable actions are promoted in addition to the installation of sustainable technology. To carry out these analyses, we used generalizable data providing evidence of the list of recurrent actions associated with the sustainable management of buildings. This approach raises the importance of standardizing the organizational process as a fundamental factor of building design to improve the control of sustainability-oriented activities in urban facilities [6].

2. Literature review: organizational vision in sustainable buildings

The Brundtland Report represents the starting point to comprehensively address the global environmental crisis. Among other important aspects, it contains the traditional definition of sustainable development, which emphasizes meeting current needs without compromising the ability of future generations to meet their own needs [7]. Environmental degradation triggered the necessity of a definition of sustainability at the time; however, due to the complexity of its transdisciplinary vision, the concept is in constant reconfiguration. Nevertheless, the role played by nature as a support, condition, and prerequisite of the production process is consistently acknowledged [8].

Although sustainable buildings can represent increased costs—from the design to the construction phase—the investment will return during the life cycle of the facility because sustainability optimizes water and energy consumption in the benefit of building users and society in general. Thus, sustainable construction is considered a holistic process to restore and maintenance the harmony between the natural and human environments by creating spaces that affirm human dignity and promote economic equity. In this context, the challenges posed by environmental problems can be considered as opportunities to adapt, change, and improve currently unsustainable practices and cease to be threats to development.

Different companies in the architectural and construction sectors are seeking to change their role as part of the problem caused by the environmental impacts associated with buildings and seek to become part of the solution [9]. Sustainable buildings certified as such under a given standard favor high value-added services and address their social responsibility with the territory by collaborating in environmental restoration. From a financial and social point of view, a sustainable building certification will bring benefits to the building's owners [10]. One definition of sustainable building involves a construction where the environment is healthy and is based on ecological principles and an efficient use of resources [11].

Organizational interaction is an essential element to achieve a sustainable construction. The organizational structure consists of defined processes that allow for an adequate execution of the activities carried out in buildings. Derived from the organizational vision and mission, sustainability should be part of its planning efforts. To achieve this, the organization must be sustainable and building users should commit to sustainable practices: sustainability is directly associated with short-term actions, planning, and projects via an organizational system that allows for the management of an important stabilizing factor for society [12]. **Table 1** presents the transcendental factors that determine the construction of an organizational system focused on sustainability.

When it comes to internalizing the need for sustainable practices in everyday life, organizational culture is possibly the central attribute, from which the rest derive. However, it can be significantly affected by the historical, political, and social context of complex and dynamic interpersonal relationships. Denison *et al.* [15] has pointed out certain conditions that influence the development of organizational practices.

The first is involvement, which corresponds to the level of acceptance of organizational goals and common values; these are the guidelines to create a balanced

Factor	Attributes
F1. Innovation and design	1.1. Construction 1.2. Environmental conditioning
F2. Organizational structure	2.1. Organizational culture
F3. Materials and resources	3.1. Green purchases
F4. Building maintenance	4.1. Building conditions 4.2. Substitution of materials 4.3. Usage patterns
F5. Cost–benefit relationship	5.1. Cost reduction 5.2. Organizational environmental performance

Source: Adapted by the authors from Liyin et al. [13] and for each factor from the following sources: F1 [10], F2 [12, 14], F3 [13], F4 [10], and F5 [13].

Table 1.
Organizational factors of sustainable facility development.

environment and improve the coexistence of active agents in their environment. It is also a practical structure that contributes to enhance organizational skills among employees. It is based on integrative learning among work teams [15, 16]. The second is adaptability, which refers to the capacity and resources of members of the organization to cope with unscheduled events. It is explained by the ability of its members to respond to unforeseen events [15, 17].

The mission states a strategic intention whose objectives are guided by the intention of shaping a future. To a certain extent, the mission articulates the vision, provides unity to the organization, and represents what is expected to be achieved in a given period [15, 18]. Finally, a climate of organizational confidence is necessary to achieve consistency. This consistency is measured by the degree to which the members of an organization have the necessary conditions for their professional development, which is consequently articulated with the relevant ethical considerations. As a result, human interactions reconcile core values that contribute to the establishment of common objectives, which have an integrating function within the organization [15].

Taking organizational life into account reveals the environment in which organizational life takes place as shown by its administrative units as it strives to increase the efficiency of its construction efforts and optimize its resources. A regulation instrument is required to monitor the users' sustainable habits. Constant monitoring and commitment is required to guarantee a timely response to the organization's internal and external issues, hence the importance of specifying the actions that contribute to sustainable practices [19].

3. Methodology: certification program categories under analysis and actions carried out to obtain certification

The purpose of the certification program created by Mexico City's Environment Secretariat in 2008 was to promote and encourage the reduction of pollutant emissions and the efficient use of natural resources based on sustainability and environmental efficiency criteria for the design and operation of buildings in Mexico City. Economic incentives are given to those who build sustainably and obtain the certification, which is valid for three years [20].

Sustainable buildings derive from the human right to have a healthy environment and the need to provide sustainable spaces for society. This right should be adequately guaranteed in the Mexican Constitution, but this is not the case; therefore, the configuration of this right must be sought in the specific laws that address it [1].

One of the pillars of such configuration is the Sustainable Buildings Certification Program (PCES), which seeks to promote these actions in order to improve habitability as a right to a healthy environment. However, a sustainable approach is not enough due to the small number of certified properties. Regrettably, sustainable buildings are still uncensused, but the program is a benchmark to join efforts in meeting the sustainability criteria established by the Environment Secretariat [21].

In this context, the analysis of this certification seeks to contribute to our knowledge about the design of criteria used to certify sustainable buildings in Mexico City. Our purpose is to examine the importance given to actions aimed at modifying user patterns to achieve sustainability in contrast with the emphasis on actions focused on the design or technological qualities of the facility. Therefore, this approach compares the two different types of factors that guide the construction of sustainable buildings. The elements of the program to be analyzed will be: 1) features of sustainable properties; and 2) criteria closely associated with organizational factors that potentially result in sustainable practices by building users (**Table 2**).

Linking the operationalization of variables based on sustainability criteria is important to enable new alternatives for urban regeneration models that allow for the association of ecological architecture with actions favoring sustainable responsibility carried out by users. Therefore, the organizational process is the driver of change, it originates the specific measures to be carried out by local agents in order to improve the constitution of sustainable buildings in the city's boroughs.

Three levels of certification are possible depending on a numeric score: compliance, between 21 and 50 points; efficiency, between 51 and 80 points, and excellence, between 81 and 100 points. The constitution of a building corresponds to its physical and technological attributes, but the present chapter includes a section on organizational factors, which are used to reveal organizational capacities and social participation toward sustainability.

The program is organized in five categories that group 47 criteria;[1] each of them has a maximum possible score to avoid all actions being concentrated in only one area. The total possible score is 120 points, distributed as follows: Energy: 40 points; Water: 25 points; Solid waste: 10 points; Quality of life and social responsibility: 25 points, Environmental impact and other impacts: 20 points. However, the score of the criteria represents a total of 221.5 points (**Table 2**).

In the present study, we classified each criteria depending on whether it was met based on the design of the facility or on the internal organization of users. This classification was used to determine the transcendence of actions aimed at modifying the users' behavior patterns.

Category	Features of the facility	Points	Organizational factors	Points
Energy	Enclosure efficiency	20	Environmental conditioning	8
	Bioclimatic design	25		
	Energy (renewable)	20		
	Solar heaters	10		
	Efficient lighting	4		
	Engines	2		
	Light control systems	2		
	High-efficiency equipment	10		
	Subtotal	*93*		*8*
Water	Rainwater collection and reuse in the facility	5	Use of efficient drinking water technologies/ savings-related elements	5
	Rainwater infiltration	5		
	Installation of wastewater treatment and reuse plants	8	Campaigns on the efficient use of water and water culture	10
	Use of wastewater treated by the municipal network	8	Leak repair	5
	Subtotal	*26*		*20*

[1] **Notes**

SEDEMA [5] states that the number of criteria is 46, but when reviewing each category, we found inconsistencies in the total number of criteria in the categories of "Energy", "Water", and "Environmental impact and other impacts".

Category	Features of the facility	Points	Organizational factors	Points
Solid waste	Infrastructure for temporary storage	3	Separation of recoverable waste	2
	Signposting	0.5	Adequate final arrangement	3
	Furniture for adequate indoors management	1.5	Dissemination and sensitization program on solid waste separation	0.5
			Special waste handling management plan (optional)	2
	Subtotal	*5*		*7.5*
Quality of life and social responsibility	Rooftop greening	7	Adequate and timely maintenance	2
	Accessibility	3		
	Ascent and descent bays for transports	1	Provision of transportation facilities to permanent users	2
	Noise level control inside buildings	2	Existence of a participatory culture concerning sustainability	2
	Bicycle parking	3	Green areas intended for comfort and to encourage social interaction	3
	Bicycle stations	2		
	Internal bike lanes	2		
	Subtotal	*20*		*9*
Environmental impact and other impacts	Parking accessibility	4	Green purchases	3
	Use of local materials	1	Elimination of refrigerants based on chlorofluorocarbons	2
	Use of materials low in volatile organic compounds (VOCs) for construction and finishing	3		
	Use of recycled construction materials	2	Indoors contamination control	3
	Recycling of existing structures	2		
	Reconversion of soil use and remediation	5	Use of biodegradable materials for the maintenance of green areas and buildings	1
	Respect for existing trees	1		
	Use of certified wood	2		
	Permeable road areas	4		
	Subtotal	*24*		*9*
	Total	**168**		**53.5**

Source: Prepared by the authors based on SEDEMA [5].

Table 2.
Classification facility features or organizational factors of the evaluation criteria used by Mexico City's Sustainable Buildings Certification Program.

Thus, the first part of this analysis focused on the criteria associated with the organization and user behavior patterns, including their assigned scores. The second focused on the actions carried out by building owners to obtain certifications, and it was based on a review of implementation resolutions [22] provided to the building owners by facility evaluators.

The present study describes such evaluation, including the actions carried out by the owners as a result of the initial inspection, which is accompanied by recommendations that must be met in order to obtain the certificate. The analyzed actions were carried out over a seven-year period (2011–2018), and we could identify favorable and unfavorable trends concerning the internal organization of users.

4. Design of Mexico City's sustainable buildings certification program and its results

The program has a number of inconsistencies in terms of total scores and required scores for certification; therefore, obtaining certification is relatively easy: only 21 points, or 17.5% of the total score, are needed for the compliance level. The category of energy alone could be enough grounds for a certification. As a consequence, the holistic nature of the assessment is lost. As has been indicated, beyond organizational actions or user behavior patterns, we focus on the importance of the criteria associated with the features of the facility.

When these criteria were classified based on the two analysis categories created for the present study, only 16 out of 47 were found to be associated with the users' organization and behavior patterns. These criteria represent 34% of the total number of criteria, and they represent only 24.2% of the total points. This difference varies depending on the category.

The most represented category is "Solid waste": 60% of the score can be obtained due to actions associated with organizational and user behavior issues. Concerning the "Energy" category, only 7.9% of the points are obtained via user actions, whereas the rest, or 92.1%, is obtained by designing and incorporating technology into the facility. In fact, in all categories except for Solid Waste, most of the points can be obtained by including technological solutions in the design of the facility (**Table 3**).

An analysis of the certification program reveals a clear preference for the use of technology in the design of a facility as the main mechanism to attain sustainability, disregarding a fundamental aspect of sustainable development, that is, sustainability as a result of the transformation of behavior patterns.

It is important to highlight the implications of such a program: the population can wrongly perceive a facility as a sustainable architectural environment only because it includes efficient technological components, and the need to modify their patterns of consumption or use of resources ecologically is never internalized. Sustainability can be hardly achieved if the population fails to take on the responsibility of using resources rationally.

A comprehensive environmental system becomes more relevant when the intention is to involve a facility's users in the improvement of the quality of their habitants. Sustainability criteria highlight the importance of aligning environmental management with current regulations to achieve substantial changes in the habitability of buildings.

Regrettably, the scope of the analyzed certification has been very limited. Despite the large number of buildings in Mexico City, only 43 of them were certified in seven years. This outcome shows that the program has ample room for improvement to design mechanisms encouraging the development of sustainable facilities. The certification classifies each facility depending on their performance status: excellence or efficiency. Half of the evaluated buildings achieved a level of excellence, which is closely associated with the total score that can be possibly obtained for the certification. The excess of points, described above, reveals one of the deficiencies in the evaluation design, which

Category	Features of the facility		Organizational factors		Total	
	Points	%	Points	%	Points	%
Energy	93	92.1	8	7.9	101	100
Water	26	56.5	20	43.5	46	100
Solid waste	5	40.0	7.5	60.0	12.5	100
Quality of life and social responsibility	20	69.0	9	31.0	29	100
Environmental impact and other impacts	24	72.7	9	27.3	33	100
Total	*168*	*75.8*	*53.5*	*24.2*	*221.5*	*100*

Source: Prepared by the authors based on SEDEMA [5].

Table 3.
Synthesis of scores by category, facility features, and organizational factors in Mexico City's Sustainable Buildings Certification Program.

Year	2011	2012	2013	2014	2015	2016	2017	2018	2019*	2020*
Buildings	3	8	8	17	13	21	11	14	5	3

Source: prepared by the authors based on SEDEMA [22].
**Certification is granted for a period of three years; therefore, the facilities that obtained it in 2017 and 2018 will have a valid certification until 2019 and 2020, respectively.*

Table 4.
Mexico City: evolution of properties registered by Mexico City's Sustainable Buildings Certification Program, 2011–2020.

makes certification easy to comply with. On the other hand, 26% of the evaluated buildings achieved an efficiency level, and 24% remained undefined in certification resolutions, but it can be inferred that they achieved compliance level.

Organizational capacity depends on actions to control and mitigate environmental contamination inside and outside the building. For that reason, the study of different types of buildings shows how specific actions have been selected to favor user dynamics and environmental comfort [23].

Most of the evaluated facilities (62%) are office buildings; this observation can be associated with the economic advantages offered by the program in terms of tax exemptions and the socially responsible image that a company acquires when it shows an interest in the environment. The remaining 29% of the facilities corresponds to condo towers that seek to reduce operational costs and provide a high-value offer to a population segment interested in environmental preservation.

Table 4 shows that the number of buildings included in the certification program increased gradually until a peak in 2016. After that year, the interest in obtaining certification has declined from 20 buildings to only three in 2020. Despite the economic benefits of certified buildings.[2] This could be possibly associated with the lack of cost–benefit advantages identified by the owners, as reflected by the lack of renewed interest in additional certifications after the initial three-year period.

[2] The two main benefits of sustainable actions are property and payroll tax breaks, an option available to private companies [5], however, these breaks must be processed independently from certification, which increases bureaucracy costs for building owners.

Document what bases the analysis [2, 5, 13, 15, 20] the documents governmental with a time-periodic publication.

The next section will analyze the actions carried out by building owners to obtain certification. This analysis was be carried based on the classification presented in the previous section and it intends to highlight the different actions aimed at strengthening sustainable organization and user behavior patterns.

5. Evolution of sustainable actions in certified buildings

The underlying design of the certification program is the main antecedent of the decisions made by facility owners to obtain the necessary scores. However, as already noted, the large number of possible points gives owners room to decide on the actions that they can easily enable depending on their specific conditions.

The present section analyzes these actions based on the classification previously presented for the certification program criteria. Again, this analysis will be analyzed by category and on a temporary basis in order to identify trends during the study period. Our aim is to determine whether these actions are linked to users' actions toward sustainability or they are limited to the use of green technologies in the design of these buildings.

As noted in the previous section, the relative weight of the criteria associated with the organizational aspects of building users was 22.3% of the total possible points. This categorically determines the type of action decided by the owners in favor of sustainability in order to obtain the certification and emphasize actions associated with the adequacy of the facility. However, as shown by graphic 2, the total number of actions associated with users' organization is higher than the proportion shown by the program design.

The proportion of actions associated with users' organization is consistently lower than the proportion associated with the features of the facility, which is 28.2% higher than the reference in all years except for 2018, when, only 27% of the registered actions concerned organizational strategies involving the users. This downward trend could be explained by the availability of economic resources that building owners can use to acquire green technologies and incorporate them into their design, which is certainly opposed to organizational measures involving the users, which until recently appeared as a feasible alternative to expensive technological solutions.

The effect of this trend can be appreciated in each of the five categories. Despite such downward trend, the Water category includes more actions focused on users' organization. Particularly, more than half of the actions carried out in 2012 were of this type. This shows the potential of water-related sustainability actions that can be transferred to the rest of the spaces in which the user interacts with the space.

In the case of the solid waste category, the strongest downward trend was observed in 2011, when 60% of the actions were associated with users' organization, but the percentage had reduced to 20% by 2018 (**Figure 1**). It is necessary to highlight this condition, since waste separation must formally be carried out by the user when generating them. Although infrastructure contributes to such activity, there are no automated systems to carry out such work. Therefore, it is essential to promote a culture of waste separation and reduction so that this category meets the objectives adequate solid waste management.

The rest of the categories are in line with the general trend, i.e., a downward trend, which emphasizes the importance of using technology in buildings over the implementation of actions that encourage changes in user behavior toward more sustainable patterns. This tendency weakens the actions in favor of favor sustainability that people can carry out and reduces accountability because people assume that the introduction of green technology will be enough to achieve sustainable objectives.

As can be observed, each building owner focuses on different actions. It is interesting to observe the territorial differentiation expressed in **Table 5**. Considering all the facilities, a total of 526 actions in favor of sustainability were identified during the period from 2011 to 2018. A third of these actions are associated with users' organization, which is a relatively high percentage in comparison with the score granted by the program's criteria.

The borough where most organization-related actions were registered was Cuajimalpa, with a percentage of 45%, followed by Cuauhtémoc, with 42%. In contrast, only 22% of the actions in certified Tlalpan properties were associated with users' organization, and only 27% in Benito Juárez. A larger number of actions associated with facility features was found in the south part of the city, as opposed to the center-west, where more actions related to users' organization were observed.

Mexico City's boroughs are quite heterogeneous; therefore, there is no simple explanation for this territorial phenomenon. For example, many of them contain

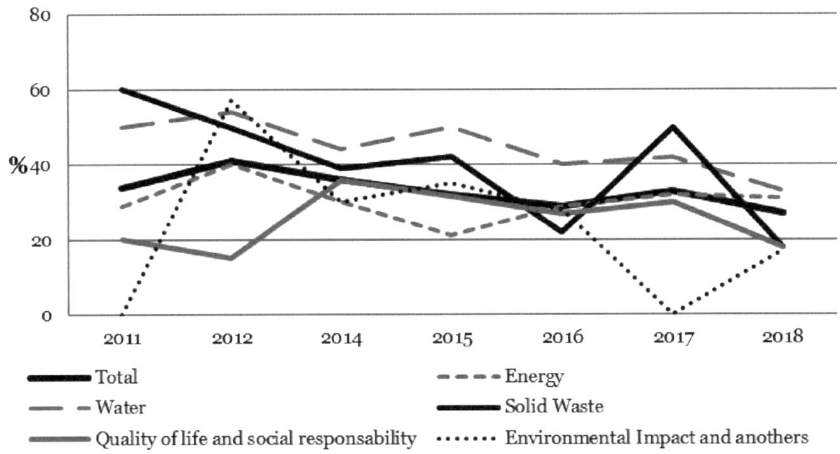

Figure 1.
Mexico City: evolution of environmental actions according to facility features and organizational factors (2011–2018). Source: Elaborated by the authors based on SEDEMA [22].

Borough	Total of facilities	Total actions	Facility features	%	Organizational factors	%	Average actions per facility
Miguel Hidalgo	13	211	145	69	66	31	16
Álvaro Obregón	8	134	91	68	43	32	17
Cuauhtémoc	5	77	45	58	32	42	15
Azcapotzalco	3	34	21	62	13	38	11
Benito Juárez	2	30	22	73	8	27	15
Cuajimalpa	2	22	12	55	10	45	11
Tlalpan	1	18	14	78	4	22	18
Total	*34*	*526*	*350*	*67*	*176*	*33*	*15*

Source: Prepared by the authors based on SEDEMA [22].

Table 5.
Mexico City: classification of actions to achieve certification.

areas of high purchasing power as well as marginalized sectors, which would require a more detailed analysis that exceeds the scope of the present chapter.

As could be expected due to the way in which the program was designed, most actions (45.1%) have to do with the energy category. However, the types of actions show a different proportion: 26% of the actions address criteria associated with energy, followed by quality of life and social responsibility, addressed in 24% of cases, which is high considering that the certification program allocates 13.5% of the score in this category to actions related to users' organization. The solid waste category also stands out; it represents only 5.6% of points of the certification, and 18% of the actions are associated with this category (**Table 6**).

Concerning differences by borough, energy-related actions predominate in four of them; however, in all cases, the proportion is lower than the score in this category; the largest proportion was found in Benito Juárez, where these actions account for 43% of the total (**Table 6**). Actions associated with environmental impact predominate in Cuajimalpa; this trend can be associated with the concern about not damaging the protected natural areas present in the borough. Finally, in Álvaro Obregón and Tlalpan, most activities were focused on quality of life and social responsibility.

Interestingly, actions tend to be heterogeneous depending on the territorial conditions of each borough, that is, depending on the most pressing issue at the local level (e.g., water distribution, energy consumption, solid waste management, social responsibility, or environmental impact). This analysis sets the tone for future studies to discuss this behavior in more detail, focusing for instance on building type; in this case, for example, the studied facilities consisted mostly of urban condos and office buildings.

On the other hand, these territory-related decisions reflect a positive attitude among building owners, who align their action plans with local sustainable goals instead of focusing on obtaining a good score and being certified. This attitude is possible because building users are made aware of the environment where they interact and act in solidarity to build their right to a healthy life. We will now broaden the analysis by discussing the five categories in more detail to identify temporal trends and actions specifically related to organizational factors.

5.1 Energy

In terms of score, this category has the highest weight on the score and on the criteria associated with the features of the facility; 92% of the score in this category can be obtained by meeting such criteria, and an average of 70% of the actions in all boroughs are focused on these features. Buildings located in Miguel Hidalgo account for 34% of the actions associated with how users organize, which is the highest percentage of all boroughs (**Table** 7).

Most reported actions concern the use of technological equipment, whereas organizational factors are clearly disregarded. Motion control and automation systems, such as hydro-pneumatic pumping, and in some cases, emergency generators and physical fire protection systems were among the most frequent technological actions. Other initiatives are energy consumption measurement programs, light and refraction control logs, use of natural lighting, solar cells, and adaptations to improve the habitability of common areas are other examples.

5.2 Water

The lack of available water resources is associated with poor consumption habits, excessive social consumption, and a lack of environmental culture. As a consequence, the recharge of aquifers is often compromised because the natural water cycle is

Borough	**Energy**	**%**	**Water**	**%**	**Solid waste**	**%**	**Quality of life and social responsibility**	**%**	**Environmental impact**	**%**	**Total actions**	**%**
Álvaro Obregón	31	23	21	16	30	22	37	28	15	11	134	100
Azcapotzalco	10	29	8	24	5	15	9	26	2	6	34	100
Benito Juárez	13	43	5	17	5	17	4	13	3	10	30	100
Cuajimalpa	0	0	6	27	4	18	4	18	8	36	22	100
Cuauhtémoc	21	27	13	17	17	22	18	23	8	10	77	100
Miguel Hidalgo	56	27	44	21	33	16	48	23	30	14	211	100
Tlalpan	4	22	2	11	3	17	6	33	3	17	18	100
Total	*135*	*26*	*99*	*19*	*97*	*18*	*126*	*24*	*69*	*13*	*526*	*100*

Source: Prepared by the authors based on SEDEMA [22].

Table 6.
Mexico City: total actions by environmental category by borough, 2011–2018.

Borough	Total actions	Features of the facility	%	Organizational factors	%
Miguel Hidalgo	56	37	66	19	34
Álvaro Obregón	31	22	71	9	29
Cuauhtémoc	21	16	76	5	24
Benito Juárez	13	9	69	4	31
Azcapotzalco	10	7	70	3	30
Tlalpan	4	3	75	1	25
Total	*135*	*94*	*70*	*41*	*30*

Source: Prepared by the authors based on SEDEMA [22].

Table 7.
Energy-related actions carried out to obtain certification in Mexico City, 2011–2018.

Borough	Total actions	Features of the facility	%	Organizational factors	%
Miguel Hidalgo	44	29	66	15	34
Álvaro Obregón	21	12	57	9	43
Cuauhtémoc	13	6	46	7	54
Azcapotzalco	8	3	38	5	63
Cuajimalpa	6	2	33	4	67
Benito Juárez	5	3	60	2	40
Tlalpan	2	1	50	1	50
Total	*99*	*56*	*57*	*43*	*43*

Source: Prepared by the authors based on SEDEMA [22].

Table 8.
Water-related actions carried out to obtain certification in Mexico City, 2011–2018.

disturbed. In this category, the possible points to be obtained due to organization-related actions represent 43.5% of the total and points related to building features represent 56.5%; therefore, this category was found to be the best balanced.

On average, the actions carried out by building owners to obtain certification are organized in practically the same way in this regard (43–56%). The cases of Azcapotzalco and Cuajimalpa stand out, since the actions carried out by building owners focused on users' organization are 63 and 67%, respectively (**Table 8**).

Integral water management policies have achieved a balance by incorporating the relationship between organizational factors and building features. Among our findings are internal campaigns carried out by companies and efficient water use programs in condos. In some cases, checks for leaks are conducted daily, and some owners have built treatment plants to irrigate green areas for use in common areas or domestic use. In addition, recent proposals have focused on distributive justice, such as rainwater collection and purification or reinstating traditional practices such as preserving natural waterways to allow for the infiltration of clean water into the soil.

5.3 Solid waste

The treatment of solid waste often requires structural changes to respond to the increasing numbers of residents and visitors in urban buildings. For that reason, infrastructure such as transfer stations and sorting, compaction, and recycling plants is currently used for processing 60% of the waste produced in Mexico City.

Borough	Total actions	Features of the facility	%	Organizational factors	%
Miguel Hidalgo	33	25	76	8	24
Álvaro Obregón	30	17	57	13	43
Cuauhtémoc	17	7	41	10	59
Azcapotzalco	5	3	60	2	40
Benito Juárez	5	3	60	2	40
Cuajimalpa	4	2	50	2	50
Tlalpan	3	3	100	0	0
Total	*97*	*60*	*62*	*37*	*38*

Source: Prepared by the authors based on SEDEMA [22].

Table 9.
Mexico City: actions associated with solid waste certification, 2011–2018.

As previously mentioned, this category has a score ratio of 60–40 in favor of actions associated with users' organization, contrary to the previous cases, where the proportion of actions of this type was higher or equal to the score. Actions associated with facility features are clearly predominant (62–38) (**Table 9**).

In Cuauhtémoc 59% of the actions are related with user's organization. By contrast, in Tlalpan, all actions are associated with facility features, which raises questions about the efficiency of waste management actions in facilities where no actions requiring the users to separate waste have been put in place.

5.4 Quality of life and social responsibility

Social involvement in cross-sectional interventions focused on environmental management will be possible provided that users become aware of the importance of their commitment as agents responsible for the dissemination of sustainable-oriented actions. Therefore, most of these actions would be expected to be associated with user's organization.

However, the proportion is very close to the score given by the certification program, that is, in which 69% of the actions are related to facility features. In other words, 71% of actions are associated with facility features, whereas only 29% are associated with users' organization. Cuauhtémoc represents a higher proportion of actions associated with activities carried out by users, accounting for 39% of the total actions (**Table 10**). In contrast, in Cuajimalpa, all actions are related with the design of the facility, which challenges the adequacy of this strategy on social responsibility.

A wide range of actions translates into better chances to improve social interaction; however, the user should find it easy to comply with sustainability-related actions. At the same time, building users should be rewarded with comfortable environments in terms of infrastructure, logistics, and enough space to engage in leisure activities. Therefore, the activities of a building's users are essential to foster social responsibility, an essential element in the transition to sustainability.

5.5 Environmental impact and other impacts

This section refers to environmental contamination and the impact of the construction itself on the environment, both of which must remain within permissible levels. The proportion of actual registered actions is similar to the score suggested by the program. Therefore, the highest possible score based on criteria associated users' organization represents only 27.3% of the total, and the remaining 72.7% represents actions associated with the design of the facility.

Accordingly, 29% of the actual actions are related with user activities. The case of Cuajimalpa stands out: half of the actions carried out in this borough involve building users. In contrast, Benito Juárez, Tlalpan, and Azcapotzalco focused their actions entirely on facility features (**Table 11**).

Environmental building safety is a result of environment-related regulations intended to guarantee the reduction of harmful and hazardous contaminants. These regulations allow suppliers to demonstrate that they are operating within current legislation and that their emissions are permissible at the local level [5].

In this regard, program design and actual actions are clearly related. The proportions are very similar in three of the five cases: water, quality of life, social responsibility, environmental impact, and other impacts. Concerning the energy category, the balance is relatively in favor of criteria associated with sustainable activities. However, most of the actions in this category are limited to facility design features. In the case of solid waste, again, most actions have to do with facility features, which demonstrates the preference for technological solutions over changing the population's behavior, an essential element to promote environmental awareness.

Borough	Total actions	Features of the facility	%	Organizational factors	%
Álvaro Obregón	37	27	73	10	27
Azcapotzalco	9	6	67	3	33
Benito Juárez	4	3	75	1	25
Cuajimalpa	4	4	100	0	0
Cuauhtémoc	18	11	61	7	39
Miguel Hidalgo	48	35	73	13	27
Tlalpan	6	4	67	2	33
Total	*126*	*90*	*71*	*36*	*29*

Source: Prepared by the authors based on SEDEMA [22].

Table 10.
Mexico City: actions associated with Quality of life and social responsibility certification, 2011–2018.

City Halls	Total actions	Features of the facility	%	Organizational factors	%
Miguel Hidalgo	30	19	63	11	37
Álvaro Obregón	15	13	87	2	13
Cuajimalpa	8	4	50	4	50
Cuauhtémoc	8	5	63	3	38
Benito Juárez	3	3	100	0	0
Tlalpan	3	3	100	0	0
Azcapotzalco	2	2	100	0	0
Total	*69*	*49*	*71*	*20*	*29*

Source: Prepared by the authors based on SEDEMA [22].

Table 11.
Mexico City: Actions for certification in the area of environmental impact and other impacts, 2011–2018.

6. Conclusions: assimilation of sustainable practices by users

Around 78% of Mexico's population is expected to be located in urban environments by 2050, ranking eighth at the international level [24]. Although agglomeration economies favor the efficient use of resources in cities, the growing demand for natural resources and infrastructure services has an undeniable impact on nature [25], and buildings play a central role in such demand.

Buildings require energy and water to operate and to enable the activities carried out by human beings who use them; therefore, they are sources of pollution [3]. The present study identified a clear trend toward sustainability based on technological solutions included into the design of the facilities. However, a lack of maintenance can reduce the efficiency of the technology or affect its performance. This is an essential factor to involve users in sustainable actions.

The practical impact of this research deal with the organizational factors is the element that adds meaning to sustainability criteria: the introduction of ecotechnology is not enough when the user will fail to use it properly. Therefore, the adoption of sustainability criteria to overhaul buildings is important, but also the incorporation of environmental culture into organizational climate, and consequently, specialized knowledge about sustainability-oriented user patterns is not enough: reasonable goals are needed to achieve habitability as a right to a healthy environment. Certification programs represent a guiding instrument for owners to conduct adequate and decisive actions in favor of sustainability. Even when buildings incorporate the latest technological sustainability features, users must become involved in the sustainable aspect of their physical environment.

The principal finding of the present study evinced the relatively low importance of organizational processes as a factor in the specific actions carried out by local agents to configure sustainable buildings in Mexico City's boroughs, and many of these actions reflect how certification programs are designed, specifically, the low consideration given to social actions.

Infrastructure and technological devices are favored in 76% of the cases, whereas only 24% of the cases favor users' organization. The proportion is very similar in terms of specific actions, and although there is a temporary trend favoring actions associated with facility features, the same proportion is observed in the program's score, since by 2018, 27% of total actions focused on users' organization issues. Improvement processes arise from the particular needs of a building, but any process must always be accompanied by constant monitoring to reproduce the intrinsic motivation of environmental awareness [26]; however, as shown by the decreasing trend in this area, true improvement may not be accomplished.

Users' willingness and abilities are essential for an adequate building management; in the present study, we differentiated the weight of each environmental item by linking them with organizational factors. Control measures are not temporary strategies: sustainable actions must be integrated into the users' activities to consolidate the environmental management program in a given building.

The actions associated with the implementation of energy-saving technologies have an outstanding presence in Mexico City's certification program; however, actual actions differed from this reference. However, the proportion is still uneven: only 30% of total actions were associated with guiding the user toward sustainable behaviors. On the opposite end of this spectrum is the category of solid waste management; in this category, the program allocates greater importance to criteria associated with users' actions (60% of the score), but building owners focused on the same proportion when implementing actions associated with the features of their facilities. The rest of the categories are in line with the score allocated by the certification program, which proves its transcendence as a guideline for owners to define the most convenient actions to obtain certification.

The value of the findings is linked with the low holistic awareness among building users is a sign of decreased interaction with the natural world, which results in poor organizational skills—users perceive themselves as separated from their environment, and they assume that sustainability will be achieved simply by using technology classified as sustainable.

The key contribution impact of this study is related with the design of Sustainable Building Certification Program. This program in Mexico City has a series of challenges, for instance, its lack of capacity to transcend into the sustainability framework of the city and its need to produce adequate guidelines to help users internalize sustainability as a vision. The first challenge is associated with the decreasing number of buildings seeking certification, which can be associated with a lack of economic incentives for the owners.

The second challenge is connected with this lack of interest; the insufficient environmental awareness among building users and owners limits their goals to economic aspects. In this regard, strengthening criteria associated with users' organization would favor the assimilation of a sustainable vision among the population and encourage the renewal of certifications under a shared goal: creating healthy environments for the population.

Mexico City's certification program is undoubtedly a sign of progress in terms of sustainability as a basic component of the urban environment. Buildings take up an average of 90% of the surface area in cities; therefore, they are the most important space in terms of urban sustainability. However, there are many areas of opportunity and strengthening the criteria to encourage sustainable actions by users is paramount.

This chapter reveals the marginal importance of the sustainable pattern users in the Sustainable Building Certification Program, nevertheless, the limitations of this research refers with the necessity to explore the central reasons of stakeholders to promote sustainable patterns in contrast with technological investment in sustainable devices. There is a clear tendency to introduce new technology as a solution to reduce environmental impact of human activity, even though the financial cost of this technology.

There is no doubt that the sustainable technology will increase and maintain its development. Future studies should analyze the manufacture environmental impact of these technologies in contrast with the investment in sustainable education and promotion of low impact patterns of urban people. We believe that the creation of sustainable consciousness will have a more effective impact in sustainability fv. Finally, users use these devices and make them efficient.

Author details

Sara Velasco-Baca[1*] and Fermín Cruz-Muñoz[2]

1 Dirección General de Estadísticas Económicas, INEGI, Mexico City, Mexico

2 Instituto Politécnico Nacional, Mexico City, Mexico

*Address all correspondence to: sarukavelbac@gmail.com

References

[1] Greaves J. Facing the Future: The Case for A Right to a Healthy Environment for Future Generations under International Law. Groningen Journal of International Law. 2020;*8*(1):30-47

[2] Comisión para la Cooperación Ambiental. *Edificación Sustentable en América del Norte*. Departamento de Comunicación y Difusión Pública del Secretariado de la CCA: Montreal; 2008. p. 80

[3] UN-HABITAT. *The Right to Adequate Housing*. Geneva: Office of the United Nations High Commissioner for human rights; 2020. p. 3

[4] Alan W, Up f J. A 50-Year Retrospective on the Implied Warranty of Habitability. Saint Louis University Law Journal. 2020;*64*(3):456

[5] Ambiente S d M, sustentables P d c d e. *Libros Blancos*. Mexico City: Gobierno del Distrito Federal; 2012. p. 90

[6] Guetterman TC, Molina-Azorin JF, Fetters MD. Virtual Special Issue on "Integration in Mixed Methods Research." Journal of Mixed Methods Research. 2020; 14, 4: 430-435.

[7] Tapia-Sisalim JD. La sostenibilidad del concepto de Desarrollo Sostenible. ¿Cómo hacerlo operativo? *Pensamiento empresarial*, 2020; *1*, 6: 184-202.

[8] Rega C. *Ecological Rationality in Spatial Planning: Concepts and Tools for Sustainable Land-Use Decisions*. Cham, Switzerland: Springer; 2020 197 p

[9] Dursun E, Varbak NS, Kılıc B. Green building certification of urban public railway transport systems for sustainable cities. Balkan Journal of Electrical and Computer Engineering. 2020;*8*(1):7-15

[10] Weiss T, Moser C, Venus D, Berggren B, Togerro A. Parametric multi-objective energy and cost analysis in the life cycle of nearly zero energy buildings – an exhaustive search approach. Sustainable Buildings. 2019;*4*:5

[11] Janik A, Ryszko A. Szafraniec M. Scientific Landscape of Smart and Sustainable Cities Literature: A Bibliometric Analysis. Sustainability. 2020;*12*(3):1-39

[12] Rosas J. *Cambio organizacional y desarrollo sustentable*. Mexico City: HESS; 2019. pp. 13-15

[13] Liyin S, Zhenyu Z, Xiaoling Z. Key factors affecting green procurement in real estate development: A China study. Journal of Cleaner Production. 2017;**153**:372-383

[14] Solís D, Robles J, Rodríguez J. Condiciones de mercado y vivienda sustentable. Vivienda y Comunidades Sustentables. 2020;7:62-76

[15] Denison H, Haaland S, Goelzer P. Corporate culture and organizational effectiveness: Are there a similar pattern around the word? In: Osland J, editor. *Advances in Global Leadership*. West Yorkshire: Emerald Group Publishing Limited; 2003. pp. 205-227

[16] Hailu A. The Influence of Organizational Culture on Organizational Performance (The Case of Debre Berhan University). Industrial Engineering Letters. 2020;*10*(1):24-35

[17] Hillmann J, Edeltraud G. Organizational Resilience: A Valuable Construct for Management Research? International Journal of Management Reviews. 2020;*0*:1-38

[18] Ogbechi D, Alase O, Taiwo F. Ogbechi ChD. Empirical Evaluation of Strategic Planning and Sustainable Business Growth Performance of Small

and Medium Enterprises (SMES). International Journal of Innovative Research and Advanced Studies. 2020;*7*(5):181-190

[19] Opoku A, Akotia J. 2020. Special issue: urban regeneration for sustainable development. Construction Economics and Building. 2020;*20*(2):1-5

[20] Secretaría del Medio Ambiente. Programa de Certificación de Edificaciones Sustentables. Gaceta Oficial de la Ciudad de México. 2008;*17*(470):8-15

[21] Secretaría del Medio Ambiente. *Inventario de Residuos Sólidos de la Ciudad de México*. Mexico City: SEDEMA. 2018. Available from: https://www.sedema.cdmx.gob.mx/programas/programa/residuos-solidos [Accessed: 2020-08-25]

[22] Secretaría del Medio Ambiente. *Dictámenes de implementación*. Mexico City: Sistema de solicitudes de información pública de la Ciudad de México; 2019. p. 91

[23] Carvalho JP, Bragança L, Mateus R. A Systematic Review of the Role of BIM in Building Sustainability Assessment Methods. Applied Sciences. 2020;*10*(13):1-25

[24] United Nations. *World Urbanization Prospects: the 2018 Revision*. New York: Department of Economic and Social Affairs, Population Division; 2019 124 p

[25] Consejo Nacional de Población. *Delimitación de Zonas Metropolitanas*. Mexico City. 2012:70

[26] Mohd R, Tangaraja G. Knowledge-sharing behaviour in public service organizations: determinants and the roles of affective commitment and normative commitment, *European Journal of Training and Development*, 2020; *ahead-of-print*, ahead-of-print.

Chapter 5

Energy-Efficient Landscape Design

Prashanti Rao and Janmejoy Gupta

Abstract

Buildings that are carefully designed using passive strategies for natural ventilation and day lighting reduces our dependency on electrical energy meanwhile ensuring thermal comfort inside the building. Similarly, carefully planned vegetation around the building helps in reducing the urban heat island effect and electricity consumption. Methodology adopted for presenting this study as book chapter, first by understanding the concept of landscape with respect to typologies and components. Secondly discussing the physical parameters in terms of temperature, precipitations and humidity of varied prevailing climatic conditions and varied methods adopted through landscape interventions and techniques to overcome the extreme conditions throughout the year, which in turn helps in reducing the consumption of energy.

Keywords: landscaping, energy-efficiency, shading, ventilation, windbreaks, climate-specific landscape design, roof-gardens, swale

1. Landscape - an introduction

Landscape is specifically the amount of land, either countryside or cityscape, that can be seen at once in a glance by the eye in a single frame. Landscape can also be referred to as an area of either land or water, taken in aggregate. According to Norman T Newton in the forward to his "Design on the Land: The Development of Landscape Architecture" book, stated landscape architecture as "the art or the science if preferred of arranging land, together with the spaces and objects upon it, for safe, efficient, healthful, pleasant human use."

Landscape elements include parks, turfs, golf courses, managed bio reserves, soil systems, water systems, street-furniture, outdoor spaces, side-walks, lighting features, railings, and of course, vegetation.

Landscape is a common element in most architectural works. It has many functional values even though it is largely used for its esthetic properties. It can be shading devices & evaporative coolers during the summer period, windbreaks during the winter period, and light filters throughout the year. Environmental quality within a building can be improved significantly by plants. Urban heat island is one of the most discussed phenomena in the present world. It is more evident in dense urban forms. It affects human beings as well as the environment by having both physical and physiological impacts. Some of the causes for this phenomenon are inappropriate material selection for building envelopes, improper land use, transportation & traffic, impermeable surfaces, etc. However, the landscape can improve this situation at different levels. Urban heat island can be reduced by proper planning of vegetation around the dwellings at micro and macro levels [1].

IntechOpen

1.1 Components of landscape

Hard-scape & soft-scape are the two constituents of landscape design. Both differ in their characteristics; Hard-scape has a solid character while soft-scape is more fluid in nature. Hard-scape remains unchanged throughout time but soft-scape changes with time as they mature. Soft-scape acts as a breathing animated component of the landscape (**Figures 1–3**).

Hard-scape elements – walkways, driveways, rocks, paver patios, etc.

Soft-scape elements - flowers, trees, turf, plants, vines, shrubs, etc.

Some of the key advantages of landscaping are as follows:

- Reducing energy consumption, CO_2 impact & heat island effect.
- Treating nitrogen pollution in rain.
- To negate acid rain effect.
- Aesthetical value

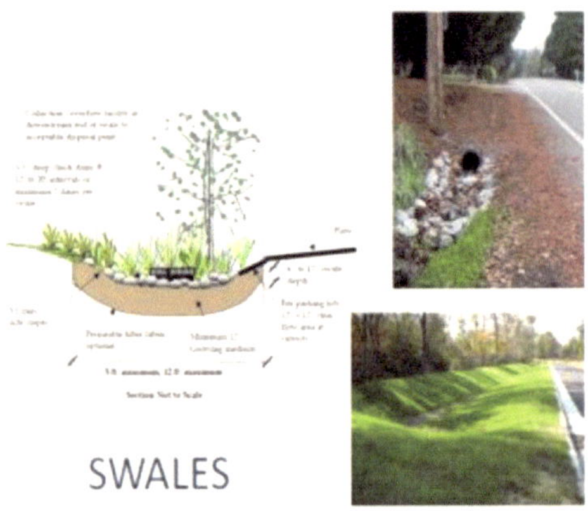

Figure 1.
Softscape (vegetated swales). Source: Co-author.

Figure 2.
Hardscape (hard paving).

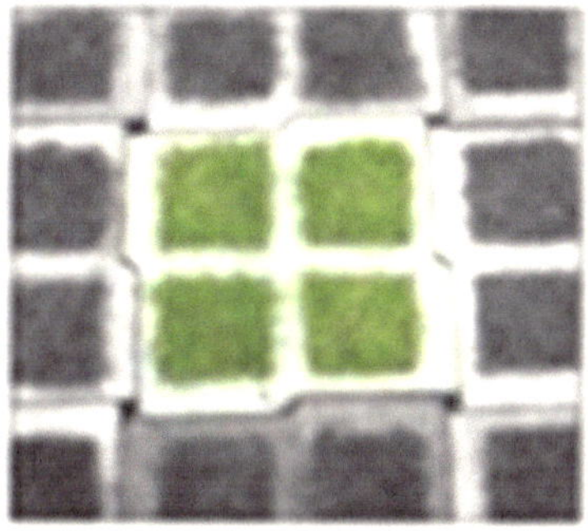

Figure 3.
Energy efficient landscape solution-combination of hard and soft paving: grass-paver blocks: Source: Co-author.

Some of the major concerns that landscape deal with are:

- Improving both outdoor & indoor environmental quality.
- Integrating man-made structures such as agriculture, forests, transport, settlements, industry etc. to the natural world.
- Composing land, water & vegetation with built elements & paving.
- Designing land, together with spaces & objects upon it, to develop a public realm.

2. Types of vegetation and its landscape-oriented benefits

Growth habits define the shape or form of the plants and play a key role in both their esthetic character and their function in the landscape. Based on growth habit, plants can be classified as trees, shrubs, groundcovers, or vines. The boundaries between these growth habit types are not always distinct, nor consistent. A plant species may fall into several categories depending upon the conditions of a particular site or its maintenance regime. Despite the shortcomings of this classification system, it is widely used in landscape design.

Herbaceous & woody plants are the main two kinds of landscape plants. Both kinds of plants differ in their characteristics. Plants that do not produce woody stems are known as herbaceous plants. They are known botanically as herbs. They can grow either in an upright, prostate, or creeper manner. Trees, shrubs, or woody vines are different types of woody plants. It is usually difficult to find out a clear distinction between trees and shrubs. Woody plants can be further divided into evergreen & deciduous plants. Deciduous trees usually shed their leaves in autumn while evergreen trees keep their leaves. Trees are relatively larger than shrubs. Shrubs are usually taller than 0.5 m and less than 3 m in height. Plants that grow over and cover ground areas are known as ground covers. They act as the bottom layer in a planting design. Groundcover has various advantages and can be used for different purposes. It can protect from drought & soil erosion. It is also used to improve the aesthetical value of a landscape as it fills the area between trees & large plants. Vines are climbing plants that can spread in different directions. Vines can be used to control erosion as well as for protecting horizontal & vertical planes from the summer sun (**Figures 4** and **5**).

Plants can also be classified according to their lifespan. Ecological origin, growth habit & seasonal pattern of plants are important factors that have to be considered in a landscape design. Region or place from which a plant species originated is considered as its ecological origin. Therefore, plants can be either native or non- native. Native plants usually have an integrated growth habit & pattern with its native ecology. Native plants can provide food to native insects and birds, developing an

Figure 4.
Trees with different foliage columnar, oval, weeping, conical and round-spread out. Source: Authors.

Figure 5.
Typical broad-leaved-evergreen tree with large canopy. Source: Authors.

Sr. no.	Name	Physical characteristics	Function/benefits
1.	Ground cover	Typically grow to less 0.5 m tall or are maintained at that height. In general, they reach 15–30 cm high.	A groundcover is utilized to provide protection from erosion and drought, and to improve the esthetic appearance of a landscape by filling areas between large plants and trees.
2.	Shrubs	It can be defined as being larger than 0.5 m, but less than 3 m in height.	Used for Esthetic purpose, Buffer and fencing also sometimes
3.	Vines/Climbers	Vines are climbing & rambling plants.	They are used on man-made structures such as a trellis, a pergola, a balcony to protect from summer sun the horizontal and vertical planes. Various vines can also be used for an effective erosion control.
4.	Trees	Trees measuring 3 to 6 m is height can be classified as `small trees`, trees 6 to 9 m can be considered as `medium trees`, and trees taller than 9 m can be considered as `large trees`.	Form and Foliage persistence of the tree perform various function, in terms of shading, wind breakers, cooling, buffering and also add Esthetic value.
4a.	Deciduous	Deciduous plants are those which completely or significantly, shed their foliage during the winter or dry season. and remain bare for a period of time, followed by the growth of new leaves in the next growing season, typically spring.	These trees are helpful for shading during the w Summer season and Allowing Sun light and warmth during the winter season.
4b.	Evergreen	Evergreen plants retain foliage throughout the year.	All of these plants have special leaves that are resistant to cold and/or moisture loss. Evergreens may continue to photosynthesize during the winter or dry period.

Source: Authors.

Table 1.
Classification of vegetation on the basis of physical characteristics and function.

inter-relation with the native ecology. They usually require comparatively lesser resources & maintenance. So that, the use of non-native plant species may not help the native ecology & also can have negative impacts in some cases.

Some plant species shed their leaves during a period of the year while some others keep their leaves throughout the year. So that annual retention of foliage is another factor that has to be considered in landscape design as this foliage retention pattern can have a larger impact on its properties such as protection from sun & wind. Landscape can implement improvements by applying principles & practices that are environmentally friendly. Some of the changes that can be made in the micro-climate of urban public green spaces, streets & housing by proper planning, implementation & management of landscape are (**Table 1**).

- Increasing the esthetic value
- Increasing the real estate value

- Decreasing the maintenance cost
- Improving energy efficiency

3. Energy-efficiency through landscaping: interventions

3.1 Introduction

Energy consumption can be reduced to a greater extent by proper utilization of landscape elements such as trees, shrubs, ground covers, or vines in strategic locations and proper quantity. Such kind of landscape systems can convert solar heat energy into moderate thermal loads. Thus, the need for mechanical cooling is reduced. Proper planning and management of landscapes help us to achieve a higher comfort level within the buildings. Landscape elements can be used to alter the microclimate around a building to regulate the heat gain in summer & heat loss in winter.

Heat exchange within a building occurs through three different processes – air infiltration, heat conduction & transmission of solar radiation. Outside air can infiltrate into the buildings through its openings in the ceilings or walls. Passage of air is also possible through cracks around doors & windows. This is the first heat exchange process. Air infiltration results in heat gain in summer and heat loss in winter. Surfaces that face wind subject to comparatively higher air pressure as the wind velocity increases and thus air enters through the openings or cracks in these surfaces. Proper planting of plants can reduce the wind velocity and thus reducing the air infiltration. Heat conduction is the second process. Heat can conduct through materials used for constructing the building. The Rate of heat conductions depends upon the insulating properties of these materials. Landscape can also reduce the heat conduction by regulating the difference between the inner & outer surfaces of the building. Landscape elements such as trees & shrubs also regulate the solar radiation receiving on the outer surfaces. Solar radiation can reduce heat loss in the winter period by increasing the temperature of outside surfaces. The Landscape system can block cold winds during the winter period to reduce conductive heat loss. Transmission of solar radiation via windows is the third process. South facing and east or west -facing glass allows an undesirable amount of solar radiation during the summer period. Glass can also heat a building in the winter period. Proper planning & planting of vegetation helps to regulate the transmission of solar radiation in different seasons. Thus landscaping & orientation on the site are two important factors that can affect the heating, cooling & lighting of a building.

- Landscape reduces air infiltration & creates air spaces adjacent to buildings. These air spaces act as insulation.
- Landscape elements can be shading devices that can reduce the total thermal heat loads on a building, especially during the summer period. Trees are better than man-made structures to provide canopy as trees do not heat up & reradiate down.
- Vegetation cools the air in contact with it by transpiration of water from the leaves and thus reducing the cooling load on buildings. It is better for the building to be surrounded by trees, rather than concrete walls.

- The advantage of using native plant species is that they are more adaptable to the local soil, climate & pathogens.
- Longwave radiations are reduced by the trees and thus regulating the natural cooling at night. Radiant cooling will be more in an open field than in a canopy.
- Vegetation can improve the quality of daylight passing through the windows and it can also moderate the light intensity & glare from the bright sky.
- Vegetated green walls are more efficient in reducing the cooling load as compared to green roofs.

Strategic designing of the landscape is required in achieving these advantages. For example, plants are more effective when they are planted adjacent to the east & west walls, as those sides are more exposed to the summer sun. The north side requires comparatively lesser shading. The selection of plant species for shading the southern windows is difficult for a building that requires winter heat.

3.2 Shading through vegetation

Proper shading of building surfaces is an effective method to reduce the undesirable thermal load, especially during periods of high-intensity solar radiation, such as the summer period of the year. The effectiveness of shade is largely dependent on canopy spread, the height of the trees, and the location of trees & shrubs within the site. One of the best methods to reduce the air temperature is by providing shade to the building roof, south-west &west facing walls & windows. This also helps in hastening early evening cooling. South-facing roof & wall surfaces have to be shaded as these surfaces receive the majority of direct sunlight when the sun is higher in the sky. Proper plants have to be provided for shading the east or west-facing surfaces as these surfaces receive direct sunlight in the morning & afternoon. Deciduous trees can be used to block the sunlight during the summer period. Sun crosses the sky at a lower angle during the winter period but proper planting of tall trees or trimming up the branches helps to achieve desirable winter sunlight. The ambient temperature around the structure as well as the indoor temperature can be reduced to some degree by shading other parts of the building & its adjacent site. The landscape design of the site is also an important tool to reduce the reflected light towards a building from surrounding surfaces (**Figure 6**).

Vine covered frames or pergolas & high bushes can also be used for shading the surfaces. One main advantage of a newly planted vine is that it can provide shade much earlier than a newly planted tree. It is an effective method to cover east and west-facing surfaces by vertical vine-covered trellis while horizontal trellis can be used on any orientation. Bushes can be used on north-facing surfaces to block the low sun (**Figure 7**).

3.3 Directing wind

Evergreen plants can be used in landscape designing to protect the cold winter winds. These plants can be used on the north, east & west sides of a building. Both evergreen trees and shrubs are used for continuous shading or to block heavy winds. Trees and shrubs with low crowns are used as an effective windbreak system that can block wind, close to the ground. Key locations, a well-designed landscape system & proper selection of plants help to reduce the total expenses for winter heating & summer cooling of a building. It can be reduced as much as twenty-five percent.

Figure 6.
Shaded alleys under canopy cover. Source: Co-author.

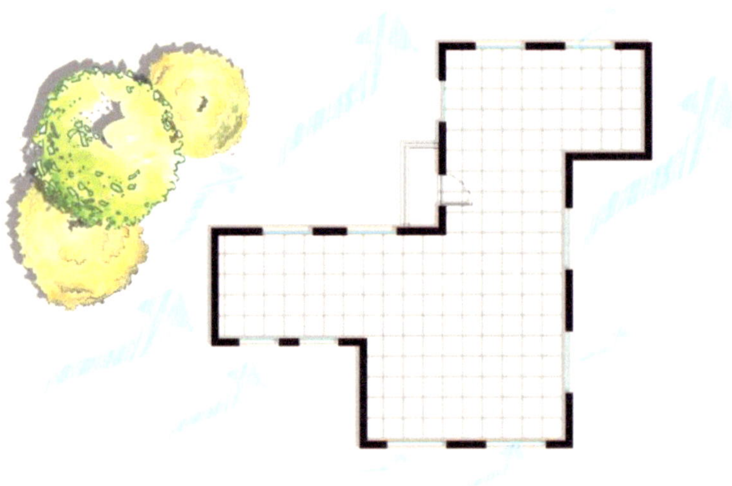

Figure 7.
Shading of west walls through landscaping. Source: Authors.

Deciduous plants are the best shading devices. They can shed their leaves in response to the change in temperature.

- Advantages - low cost & aesthetical value, glare can be reduced, these plants have the ability to cool the air via transpiration.
- Disadvantages - limited height, diseases can affect growth & slow growth

The east, south-west & south-east sides of a building are usually considered as proper locations for deciduous plants. Deciduous plants may also cause more harm during winter than good in the summer if those plants are not carefully placed on the southern side of a building.

3.4 Windbreaks

The Wind is an important climatic factor that has to be considered in a landscape design. The Windbreak is an effective system used in a landscape to control

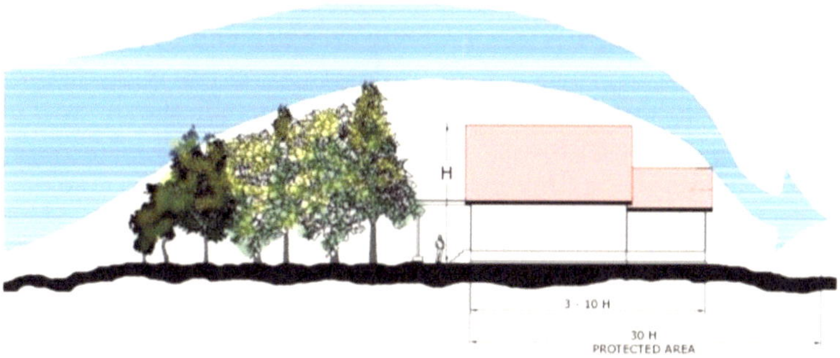

Figure 8.
Wind-shadows created by landscaping. Source: Co-author.

wind & its impacts. These shelter belts, formed by rows of trees and shrubs can reduce the wind speed or redirect its movement. Properly planted windbreaks in a cold climate can regulate the heat loss & air infiltration by reducing the wind velocity around the building. Plants that can withstand the winter climate conditions have to be selected for windbreak systems. It is an effective technique to use evergreen species as the major constituent of a windbreak composition. They occupy a significant portion of the system as these plant species retain wind-blocking mass in winter. These species can divert cold winds from the buildings & thus reducing the expense for heating. Distance between the tree and building depends upon the tree height. The Optimum distance is usually taken as 1 to 3 times the windbreak height. Distance up to which the wind can be reduced depends upon the height of the tallest row. It is considered to be 30 times the height. The effective distance of a windbreak system is generally indicated based on the windbreak multiplier, which is measured from the middle of the outermost layer of vegetation, downwind, along a line following wind direction. Components of a windbreak system have to be closely spaced to act as a continuous barrier against the wind. It is necessary to consider the final form that plant species will achieve once they get matured. A one-row windbreak system is formed by a single linear row of trees & shrubs. Evergreen plant species are densely planted in a one-row windbreak as these species retain their lower limbs & foliage. Deciduous plants with narrow clowns can be used and these plant species have to be densely planted. Two-row & twin-row windbreaks are composed of two linear rows of trees or shrubs. A single species, a set of 2 species, or a mixture of species can be used to form this windbreak system. Each row of the windbreak has to be densely planted in the same way as in a one-row windbreak. Three-row windbreaks are composed of 3 rows of trees or shrubs. It should include at least one row of dense evergreen trees. Other rows can be either deciduous or evergreen plantings. Shrubs can also be used as a front-row to catch the snow, if necessary. The three-row system provides additional sheltered spaces and there is a possibility for greater diversity. Thus, the three-row windbreak system has more wildlife value than a single or double-row system (**Figure 8**).

4. Energy-efficiency through landscaping - techniques for different climatic zones

The mature size of plant species has to be considered in designing the landscape. The growth rate is thus an important factor. Fast-growing plant species may not be a good choice as most of these species have poor strength even though some vines are effective fast-growing species that can be used in landscaping. Man-made structures such as a pergola, framework, wall, etc. can be used for supporting vines. Vines can also act as shading devices. Properly designed and executed landscape consisting of trees, shrubs, vines & man-made structures can regulate the micro-climate around a building and thus reducing the heat gain in the summer and heat loss in the winter. Vegetation can guard buildings from the cold winds during winter. It also provides shade from the summer sun and controls the solar radiation during different periods of the year. Thus, landscape strategies can be used for solar, thermal & wind control, according to the climate types.

4.1 Hot and humid climate strategies

The climate of hot-humid zones is characterized by high rainfall and high humidity. The temperature range is relatively high at around 30–35°C and is fairly even during the day and throughout the year. Due to minimal temperature differences, winds are light or even non-existent for longer periods. However, heavy precipitation and storms occur frequently.

Landscape design strategies can be used to maximize the shade throughout the year and improving the air movement, thus increasing thermal comfort. Deciduous vines covering the wide trellises on the north & south sides of a building act as shading devices. These can provide comfortable outdoor areas and solar protection. Planting beds that require frequent watering have to be avoided in areas adjacent to the building. Plant species that allow penetration of low-angle winter sun have to be used for shading the buildings and outer spaces (**Figure 9**).

Deciduous trees with high-canopy can be used on the east & west sides of a building to improve its solar protection in the morning & afternoon. These trees allow the movement of air underneath the canopies. Low vegetation has to be kept away from the building. This allows breezes to pass through and also prevents dampness. Proper landscape placements & deflection techniques help to channel prevailing winds and thus increasing the air movement. Glare & heat absorption can be reduced by light-colored materials that are paved around the building (**Figure 10**).

Other strategies which shall be adopted to minimize the discomfort occurred due to hot and humid climatic conditions like buildings should be separated with large, free spaces between them. This allows airflow which provides ventilation for cooling and a hygienic environment. Certain species of trees (e.g. rain trees) form

Figure 9.
In buildings it is possible to have vines/creeper covered trellises which provides shading in summer, allows diffused sunlight. Source: Created by Nivedhitha Ravichandran.

Figure 10.
Channeling prevailing winds with appropriate landscaping. Source: Authors.

an extraordinary outdoor space by creating a canopy effect. They should not be planted too far from each other, so that the crowns form a wide hall-like space, creating a comfortable microclimate. An un-shaded pavement should be avoided as far as possible and air should not be allowed to pass over such hot surfaces before reaching buildings. High bushes, however, should be avoided near buildings because the space between the ground vegetation and the high crowns of the trees should remain open, providing free access for the wind at the level of the living spaces.

4.2 Hot and dry climate strategies

The hot and dry climate, it is imperative to control solar radiation and movement of hot winds. The design criteria should therefore aim at resisting heat gain by providing shading, reducing exposed area, controlling and scheduling ventilation, and increasing thermal capacity. Hence, strategy should be such that it should prevent formation of dry dusty air due to overheating. It should maximize filtered air movement in summer. Vegetation is desirable as a radiation absorbent surface and for it has evaporative and shade giving properties.

The main aim of landscape design should be to provide maximum shade during the late morning & late afternoon hours. Forestation can be avoided on the north & south sides of the building and landscape elements such as shrubs, deciduous trees, vines, etc. can be used at the eastern & western sides. More shade-providing trees can be used at the east & west sides of the building. This helps to improve shade. Solar heating of the southern walls can be regulated by using shade trees or trellis structures with vines.

Vines help to cool the air adjacent to it via transpiration. Vines growing on vertical structures can also protect the east & west sides from heat gain during the morning & afternoon. Water features are also effective landscape tools that can cool the air in a hot dry zone. The Cooling effect is produced when hot, dry winds pass across the water body. This generates required moisture. It is better to reduce the use of paving materials and provide vegetation as much as possible. This reduces the glare as well as the potential for heat absorption by the paving materials. Light-colored paving material is an effective choice. Courtyard & garden walls are the other tools that can keep out the hot winds and conserve moist air. By planning narrow winding alleys and streets, which are shaded and relatively cool and break stormy winds, but allow through-ventilation and adequate natural lighting.

4.3 Composite climate strategies

The Composite climate is characterized by three seasons. A hot and dry season, usually the longest period, is followed by a wet and warm season, the monsoon period. In the third season, the winter time, depending on the altitude, temperatures can drop far below the comfort level, especially at night, whereas daytime temperatures are moderate and the solar radiation intense.

It is necessary to maximize the shade throughout the year as well as improving the air movement. Proper humidity levels have to be maintained in dry seasons. These are the main objectives for landscape design in a composite climate zone. Some of the landscape design interventions that should be used in a composite climate zone are as flows:

- Water features are effective strategies for cooling effect during dry seasons in a composite climate zone.

- Deciduous trees with high canopy & terrace gardens can be used on the west & east sides of a building. This provides solar protection during morning & afternoon. It also allows air movement underneath the canopies.

- Light-colored paving materials can be used around the building as this reduces the glare & heat absorption. Channeling prevailing winds with wind channeling & deflection techniques helps to maximize the air movement.

Wide trellises with deciduous vines can be used as shade structures on the north & south sides of a building to provide additional help for solar protection. It also develops comfortable outdoor spaces (**Figures 11** and **12**).

4.4 Temperate climate strategies

Temperate climates are generally defined as environments with moderate rainfall spread across the year or portion of the year with sporadic drought, mild to warm summers and cool to cold winters.

It is necessary to consider more substantial seasonal variations to effectively accommodate the climatic conditions of temperate regions. It is a good method to increase shade during the summer and the warming effect of the sun in winter. Winter winds should be prevented while summer winds have to be directed towards the buildings. Some of the elements that can be used in a prototypical landscape design for these climate regions are:

- Use of high-canopy deciduous trees with high branches on the west & east sides. This allows penetrating warming rays from the lower sun in the winter as well as it protects from the high summer sun.

- Use of low-branching evergreen tree clusters to block cold winds from north-west or north-east during the winter.

- Dense evergreen shrubs can be used on the north, west, and east sides of a building to create an insulating air space between the building & vegetation. This helps to reduce heat loss during the winter.

- Distance between the building and windbreak on the north side should not be more than four times the height of the windbreak

Figure 11.
Window shaded by trellises with deciduous vines. Source: Co-author.

Figure 12.
Deciduous trees cutting off summer solar radiation. Source: Co-author.

- An overhead trellis with deciduous vines can be provided adjacent to the southern façade as it adds additional shade to the building. It also creates a shaded outer space for summer use.

- Light-colored outdoor paving materials reduce the heat absorption and cooler air temperature can be maintained during the warm weather climate. It also reduces glare.

- It is necessary to channel prevailing winds with wind channeling & deflection techniques.

- Low vegetation has to be kept away from the building to prevent dampness & allow breezes to pass through.

4.5 Very cold climate strategies

Regions that lie in the cold climate zone are situated at high altitudes. The temperatures range between 20 and 30°C in summers, while in winters, it can range from −3°C to 8°C, or less.

Landscape design in a cold climate region has to consider the protection of the building from norther winds in winter. Windbreaks can be densely planted to prevent these cold winter winds. Overheating from the direct summer sun can be a problem and this can be avoided by providing shade to the south & west surfaces of the building. Dense evergreen shrubs can be planted at the northern sides of the building to create dead air spaces. This acts as insulation during the winter and summer. The speed of cold winter wind can be regulated by planting dense rows of evergreen trees and forming an earthen berm on the north & northwest sides. Low shrubs & grass should be applied in the southern windbreaker. Deciduous trees can be used in the south-west and south-east directions away from the building.

Earth sheltering is also an effective landscape tool in cold climates. It can be used if the building site is located on a south-facing slope that receives sufficient sunlight. Deciduous trees and shrubs can be used on the southern side of a building, as it helps to provide summer shading when required. It also allows low winter rays.

Advantages of using deciduous high canopy trees on the east & west sides are:

- It allows warm winter rays, provides summer shade, and maximize summer breezes.

A sunken terrace with light-colored reflective material can be incorporated into the southern side of the structure to further capture low winter sun & reflect its warmth to building interiors. Darker paving materials can be used around the building to capture warmth & promote snowmelt.

4.6 Roof garden - an energy-efficient landscaping technique

Lightweight soil medium is used in green roof gardens (**Figure 13**). A drainage layer & a high-quality impermeable membrane is provided beneath the soil medium to prevent the seepage of water. Plant species are selected that can withstand severe, dry roof temperature and resist short bursts from heavy rains [2].

4.7 Swales - an energy-efficient landscaping technique

Swales are constructed wetland systems. They are mainly used for managing storm-water runoff. These systems are used to maximize the removal of pollutants from the storm-water runoff and it is carried out through settling, uptake & filtering by proper vegetation planted. Some of the advantages of using swales are [3]:

- reduce peak runoff rates
- help in rainwater harvesting efficiency
- contribute to the green goals of the site & project

5. Conclusion

Plants are immensely useful in the heating, cooling, and lighting of buildings. Simple strategies utilizing landscape planting elements such as trees, shrubs,

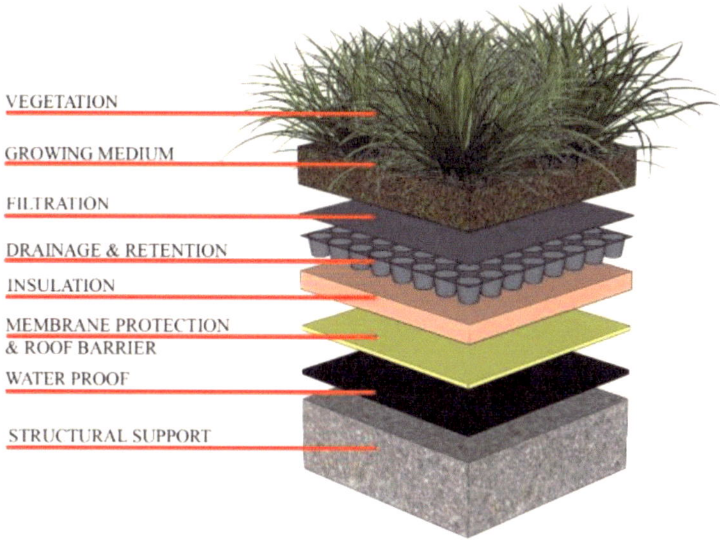

Figure 13.
Schematic cross section of a green roof. Source: Redrawn by Co-author, reference from American Wick Drain: https://www.awd-usa.com/drainage-applications/green-roof.

groundcovers or vines in key locations and in proper quantities can greatly reduce energy consumption. In certain circumstances, carefully positioned trees and shrubs can save up to 25% of the energy a typical household uses. Appropriately utilized landscape elements and systems can deflect and diffuse sunlight or dissipate solar heat energy to moderate thermal loads and reduce requirements for mechanical cooling. Energy efficient landscaping has additional benefits such as lower maintenance costs, a reduction in water use, a cleaner air.

Author details

Prashanti Rao and Janmejoy Gupta*
Department of Architecture, School of Planning and Architecture, Vijayawada, India

*Address all correspondence to: janmejoygupta@spav.ac.in

References

[1] Newton, 1974, Design on the land – The development of Landscape Architecture, The Development of Landscape Architecture (Belknap Press) Hardcover

[2] Para No. 2, American Wick Drain, accessed from https://www.awd-usa.com/drainage-applications/green-roof, August 2020

[3] Lechner & Lefroy, 2015, Greater Hunter landscape connectivity assessment. University of Tasmania, Hobart, Tasmania

Chapter 6

Towards Innovative and Sustainable Construction of Architectural Structures by Employing Self-Consolidating Concrete Reinforced with Polypropylene Fibers

Wael Zatar and Hai Nguyen

Abstract

Self-consolidating concrete (SCC) has been successfully employed to reduce construction time and enhance the quality, performance, and esthetic appearance of concrete structures. This research aimed at developing environmentally friendly fiber-reinforced concrete (FRC) consisting of SCC and recycled polypropylene (PP) fibers for sustainable construction of city buildings and transportation infrastructure. The addition of the PP fibers to SCC helps reducing shrinkage cracks and providing enhanced mechanical properties, durability, and ductility of the concrete materials. Several mix designs of self-consolidating fiber-reinforced concrete (SCFRC) were experimentally examined. Material and esthetic properties of the SCFRC mixtures that include micro silica, fly ash, and PP fibers were evaluated. Trial-and-adjustment method was employed to obtain practically optimum SCFRC mixtures, mixtures that are affordable and easy to make possessing enhanced compressive strength and esthetic properties. Slump flow and air content testing methods were used to determine the fresh properties of the SCFRC mixtures, and the esthetic properties of the mixtures were also evaluated. The hardened properties of the SCFRC mixtures were examined using three- and seven-day compression tests. The amount of fine/coarse aggregate, water, and other admixtures were varied while the Portland cement content in all mixtures was maintained unchanged. The maximum three-day compressive strength was 43.17 MPa and the largest slump flow was 736.6 mm. Test results showed enhanced material properties such as slump flow, air content and compressive strength values of the SCFRC mixtures and their excellent esthetic appearance. The favorable seven-day compressive strength of the SCFRC mixture, with 4.8 percent air content and 660.4 mm slump flow, is 39.26 MPa. The mixtures' in this study are proven to be advantageous for potential SCFRC applications in architectural structures including building façades and esthetically-pleasing bridges.

Keywords: self-consolidating concrete (SCC), fiber-reinforced concrete (FRC), mix designs, enhanced material properties, esthetic appearance

1. Introduction

Architectural applications often require concrete components with defect-free surfaces, colorful and esthetic appearance (e.g., beautiful appearances of concrete structures made of architectural colored or white concrete), and good compressive strength. Self-consolidating concrete (SCC) appears to be an excellent candidate for these applications. SCC is a highly flowable concrete that can flow under its own weight and attain good consolidation without using any external vibration for compaction and in the absence of segregation and bleeding [1]. SCC was first developed by researchers at the University of Tokyo, Japan in 1988 to obtain sustainable concrete structures [2–5]. Since then, SCC has been applied for construction of building, bridge, and tunnel throughout Japan, Europe, United States, and the rest of the world [6–8].

The term fiber-reinforced concrete (FRC) often refers to a concrete containing dispersed randomly oriented fibers [9]. FRC was first developed in early 1960s that brought it to the attention of both academia and industry around the world [10]. The inclusion of fibers (e.g., steel, polypropylene, polyvinyl chloride, nylon, carbon, glass, etc.) into the mix design of FRC can enhance mechanical properties and ductility of concrete though they can adversely affect workability of FRC mixtures. In this study, polypropylene (PP) fibers were used because they can provide a restrained shrinkage cracking and an improved spalling resistance for concrete (when the PP fibers melt, they provide channels that allow evacuating the vapor and relieving the internal pressure to avoid explosive spalling and loss of concrete cross-section in fire) [11].

Mix-design method plays critical roles in mechanical properties of SCC, which largely affected by the characteristics of constituent materials and mix proportions. The workability, passing and filling abilities, and segregation resistance are important characteristics of well-designed SCC mixtures. The filling ability of SCC (a.k.a. unconfined flowability or deformability) can be defined as the ability of concrete to flow into and fill in all spaces of the formwork under its own weight. The passing ability, on the other hand, is the ability to flow through tight openings under its own weight [12]. ACI committee 237 [13] developed standard test methods to measure the main characteristic of SCC, as shown in **Table 1**. Slump flow, V-shaped funnel, Orimet, and T50 are common methods to characterize the filling ability of SCC. The J-ring test method is used to determine the passing ability of SCC through reinforcing steel [14]. L-box and U-box tests are used to characterize both the passing and

Test method	Category	Characteristic	Measurement
Slump flow	Free flow	Filling ability	Flow distance
V-shaped funnel and Orimet	Confined flow	Filling ability	Flow rate
T_{50}	Free flow	Filling ability	Flow rate
J-ring	Confined flow	Passing ability	Flow rate
L-box	Confined flow	Passing and filling abilities	Flow rate and distance
U-box	Confined flow	Passing and filling abilities	
Visual stability index (VSI)	Static condition	Segregation resistance	Visual stability of the mixture
Column segregation test	Confined flow	Segregation resistance	Segregation of aggregates

Table 1.
Standard test methods for SCC (reproduced from ACI 237R-07 [13]).

filling abilities of SCC. The visual stability index (VSI) and column segregation test methods are developed to assess the segregation resistance characteristics of SCC. The VSI test involves the visual examination of the SCC slump flow spread and is used to evaluate the relative stability of SCC mixtures. Highly stable batches (VSI value = 0) show no evidence of segregation in slump flow spread while highly unstable batches (VSI value = 3) clearly indicate large segregation (i.e., large aggregate pile in the center of the concrete spread and/or large mortar halo). The VSI values for stable and unstable SCC batches are 1 and 2, respectively [13, 15].

Su et al. (2001) [16] proposed a new mix design method, which was economical, less time-consuming, and simpler than the traditional Japanese Ready-Mixed Concrete Association. The amount of aggregates, water/cement ratio, and type/dosage of superplasticizer were proved to significantly affect the properties of SCC. High quality SCC can be produced through trial mixtures and tests including slump flow, V-funnel, L- and U- boxes, compressive strength tests. The packing factor of aggregate (the aggregate content), defined as the ratio of mass aggregate in tightly pack state to that of loosely packed state, influenced the strength, flowability, and self-compacting ability of SCC. The passing ability through narrow spaces of rebars was enhanced because the designed SCC mixtures contained less coarse aggregates and more sand. The volume of sand to mortar was approximately 54–60%. The cement contents used in the proposed mix design method was smaller than that required by other mix design methods (because of the increased amount of sand), which result in a cost saving.

Sonebi et al. (2007) [17] carried out a composite factorial design study to investigate the influence of three key parameters of mixture composition on filling and passing abilities of SCC. The examined parameters were the dosages of water and high-range water-reducing admixture (hereafter called HRWR), and the volume of coarse aggregates. Slump flow, T50, T60, V-funnel flow time, Orimet flow time, and the blocking ratio (L-box) were employed to measure both passing and filling abilities. The results revealed that dosages of water and HRWR and the volume of coarse aggregate significantly affected the slump flow at 5, 30, and 60 minutes. T50 tests showed an insignificant influence of all investigated parameters including dosages of water and HRWR and volume of coarse aggregate. On the other hand, T60 test exhibited a significant influence of dosages of water and HRWR. The L-box blocking ratio was found to be substantially affected by dosages of water and HRWR and the volume of coarse aggregate. The L-box test was proved to be suitable for assessment of the passing ability of SCC through closely spaced rebars.

Felekoğlu et al. (2007) [18] investigate the effects of water/cement ratio and dosages of superplasticizer on the fresh and hardened properties of SCC. Slump flow, V-funnel, and L-box tests were carried out to determine optimum SCC mixtures. The results showed that optimum water/cement ratios for best SCC mixtures were in the range of 0.84–1.07 by volume. The ratios beyond this range may result in segregation or blocking of the mixtures. Trial-and-error method was recommended to produce proper concrete mixtures. Higher splitting tensile strength and lower modulus of elasticities were obtained from the SCC mixtures compared to conventional vibrating concrete.

Uysal and Sumer (2011) [19] presented experimental investigations on the performance of SCC. Conventional Portland cement was substituted with fly ash, granulated blast furnace slag (GBFS), and limestone/basalt/marble powders. The compressive strength, density, ultrasonic pulse velocity, and sulphate resistance of SCC were characterized by the amount of mineral admixtures. The results revealed that the workability and compressive strength of SCC mixtures were significantly increased with the addition of the fly ash and GBFS mineral admixtures. The compressive strength reached 105 MPa at 400 days when substituting 25% of Portland cement with the fly ash. The added mineral admixtures yielded favorable effects on the strength loss because of sodium and magnesium sulphate attack.

Recently, colored SCC (C-SCC) for architectural concrete has been developed [20–22]. López et al. (2009) [20] conducted an experimental investigation on C-SCC and presented advantages of using a mortar-based mix design method for C-SCC. Determination of the optimum proportions of SCC mixtures (consisting of various types of pigments, cements, and mineral/chemical admixtures) were quick and easy by using the mortar-based approach. The colorimetric parameters, color homogeneity, surface finishing, and the effects of pigments on the viscosity of C-SCC were effectively evaluated with this approach.

2. Research significance

The aims of this study are two folds: (1) To develop optimum SCC mixtures using trial-and-error method; and (2) To create architectural concrete with excellent surface finish (i.e., free of honeycombing or signs of discoloration and bleeding). First, several trial batches were conducted to obtain optimum SCC mixtures with adequate balance between filling and passing abilities. An experimental program was implemented to investigate the influence of the key mixture parameters (e.g., the volume of fine and coarse aggregate, dosage of water, high-range water reducing admixture, and viscosity-modifying admixture) to the material properties of self-consolidating fiber-reinforced concrete (SCFRC). Second, several mix designs of SCFRC were tested to achieve finest characteristics of architectural concrete (e.g., smooth appearance, defect-free surface, reduced cracks, improved durability, and good compressive strength). With the use of environmentally friendly products in SCFRC (i.e., recycled PP fibers and mineral admixtures such as fly ash and micro silica), the developed SCFRC can provide excellent environmental and esthetic solutions for sustainable construction of architectural structures such as bridges and buildings.

3. Experimental program

3.1 Materials and methods

Portland cement used for this study conformed to ASTM C150 [23]. Coarse aggregates with a maximum size of 12.5 mm were conforming to ASTM C33 [24]. Supplementary cementitious materials, such as fly ash and micro silica (a.k.a. micro silica fume) were added to the SCFRC mixtures to improve the workability, strength, and durability. Fly ash added to the SCFRC mixtures was conforming to ASTM C618 [25]. Micro silica was used for the SCFRC mixtures to enhance segregation and bleeding resistance. The HRWR admixture conformed to ASTM C1017 [26]. Superplasticizer was added to the SCFRC mixtures to enhance the flowability of concrete. In addition, viscosity-modifying admixture (VMA) was used to improve the stability of the SCFRC mixtures. The amount of VMA was adjusted after each SCFRC batch to obtain the designed level of stability. Furthermore, Air-Entraining Admixture (AEA), conformed to ASTM C260 [27], was used for the SCFRC mixtures to improve the freeze–thaw resistance of the concrete.

Polypropylene (PP) fibers were added to the SCFRC mixtures (0.25 to 0.5 percent by volume of concrete) to prevent concrete cracking due to plastic shrinkage. The use of PP fibers in the SCFRC mixtures has advantages including no water demand and high resistance to chemical attack. The material properties of the PP fibers are presented elsewhere [28]. Mixture proportions for 14 selected batches of the SCFRC mixtures (per 1/20 yd^3 or 0.038 m^3) are shown in **Table 2**.

Batch	Cement (kg)	Fly ash (kg)	Micro Silica (kg)	Sand (kg)	Gravel (kg)	Fibers (kg)	Water (kg)	AEA (ml)	HRWR (ml)	VMA (ml)
1	17	4.54	1.09	28.94	30.07	0.14	8.16	2.3	125.0	5.0
2	17	4.54	1.09	27.35	28.49	0.14	7.03	3.0	125.0	5.0
3	17	4.54	1.09	25.99	30.89	0.14	8.16	4.0	155.3	10.4
4	17	4.54	1.09	25.99	30.89	0.14	8.16	3.5	155.3	10.4
5	17	4.54	1.09	25.99	30.89	0.14	6.80	3.5	165.0	10.4
6	17	3.45	2.18	25.99	30.89	0.14	7.48	3.0	165.0	10.4
7	17	4.54	1.09	25.99	30.89	0.14	7.48	1.8	165.0	10.4
8	17	4.54	1.09	25.99	30.89	0.14	7.48	5.9	165.0	14.8
9	17	4.54	1.09	25.99	30.89	0.14	7.48	3.5	170.0	20.0
10	17	4.54	1.09	25.99	30.89	0.14	7.48	3.5	165.0	20.7
11	17	4.54	1.09	25.99	30.89	0.14	7.48	3.5	180.0	20.0
12	17	4.54	1.09	38.96	46.36	0.14	12.25	3.5	255.0	50.0
13	17	4.54	1.09	38.96	46.36	0.14	12.25	3.5	248.4	31.1
14	17	4.54	1.09	25.99	30.89	0.14	7.94	3.5	180.4	20.7

Table 2.
Mix proportions of SCFRC (per 0.038 m^3 or 1/20 yd^3).

3.2 Mixing procedure

First, the ingredients (i.e., Portland cement, sand, coarse aggregate, water, fiber, and mineral/chemical admixtures) were weighed to the specifications of the mix design (**Table 1**). The air entrainer was placed into the sand before mixing was started. The gravel and air entrained sand were added to the mixer first. Then, three-quarters of the total water was added in order to saturate the aggregate and form slurry. The remaining water was used for saturation of the cementitious materials. Once the aggregate was saturated, the micro silica was added along with one-third of the remaining water. Once the micro silica was saturated and mixed into the slurry, the cement was added with another one-third of the remaining water. Then, the fly ash was added along with the remaining one-third of the water. At this point, all aggregate, cementitious material, and water was in the mixer. Once the ingredients were well mixed, the fibers were added. Once the fibers were well distributed throughout the concrete, the superplasticizer was added and left to mix for approximately five minutes. Once the superplasticizer took effect, the viscosity-modifying admixture was added and allowed to mix for five minutes before tests were performed.

3.3 Mix designs and specimen preparations

In order to ensure the quality of our concrete, tests were performed to test the compressive strength, and sample molds were poured to test the ability of the concrete to mold to the forms. As mentioned above, inverted slump tests and air content tests were also performed. The target range for the slump of the fresh concrete is 50.8 to 63.5 centimeters. The target range for air content in the fresh concrete is 3 to 5%. The target 28-day compression strength of the cured concrete is 55.2 to 68.9 MPa.

Because of the unique characteristics of SCC, new testing techniques must be used. The slump flow test measures the relative flow ability using an inverted slump cone. The slump cone was filled on a flat surface with no rodding or other consolidation methods and then lifted from the floor allowing the concrete to flow outward. The diameter of the concrete was measured in various directions in order to obtain an average value for the slump. The edge of the slump was inspected for segregation. If segregation was present, a halo of segregated water would be observed. The traditional compression testing procedures were followed except for in the preparation of the specimens. The specimens were prepared by filling the cylinders without rodding. The only other test performed was the air content by pressure method. No alterations were made to the traditional air content testing procedures. In addition to these tests, a sample of concrete was poured into a rubber mold that simulates the inside of the column form to assess the esthetic qualities of the mix.

4. Results and discussions

4.1 SCFRC mixtures

As can be seen in **Table 2**, for the first batch, the mix proportions of SCFRC (per 0.038 m^3) included 2.3 ml of AEA, 125 ml of HRWR, and 5 ml of VMA. The amount of gravel was set to 54 percent of the total aggregate to achieve a higher compressive strength. The water/cement ratio was 0.36 to attain complete hydration while reducing the water content to gain adequate strength. Batch 1 exhibited acceptable slump flow, air content, and compressive strength.

In batch 2, the AEA was increased from 2.3 ml to 3.0 ml, the water was reduced from 8.16 kg to 7.03 kg to gain more strength. The sand and gravel was reduced to

27.35 kg and 28.49 kg, respectively. The rest of the mix proportions was identical to batch 1. Batch 2 exhibited a total slump flow reduction of 88.9 mm (i.e., the slump flow diameter of batch 2 was approximately 508 mm). The air content was lower than batch 1 with an additional 0.7 ml of AEA, which may result from a quicker mix time than batch 1. The reduction of water resulted in a decreased slump flow without affecting the compressive strength.

It is worth mentioning that the performance of the SCFRC mixtures in batch 5 was close to ideal with a slump flow of 584.2 mm and an amount of air content of six percent. The three-day compressive strength of batch 5 was highest among all the tested mixtures. This batch exhibited the least desirable esthetic characteristics of all the mixtures. Based on the results from previous batches, the water in batch 14 was reduced to 7.94 kg to improve strength. The superplasticizer was increased to 180.4 ml to achieve a larger diameter slump flow and increased workability. The test results from this batch showed a good slump flow, air content, and compression values with an acceptable esthetic appearance.

4.2 Slump flow and air content

Figure 1 exhibits the experimental results from the slump flow test. The slump flow diameters of all the batches ranged between 508 mm and 736.6 mm, which was in line with the recommendation of Nagataki and Fujiwara [29] for high flowable concrete. The slump flow varied significantly from batch to batch. The record low of slump flow was 508 mm for the SCFRC mixtures in batch 2. The reason for this low might be due to the reduction of water in the mixtures while other ingredients remained constant. A maximum slump flow of 736.6 mm was found in batch 12 and the increase of slump flow may be attributed to the fact that VMA was increased while other mixing components were kept constant.

Test results of the air content tests are presented in **Figure 2**. As can be seen from the figure, the air content of the SCFRC mixtures varied in the range of 5 ±2 percent. The wide range of air content is due to the large amount of HRWR used in the SCFRC mixtures. The air content of the SCFRC mixtures in this study, however, was within the expected range. It is worth mentioning that a maximum air content of 6.2 percent

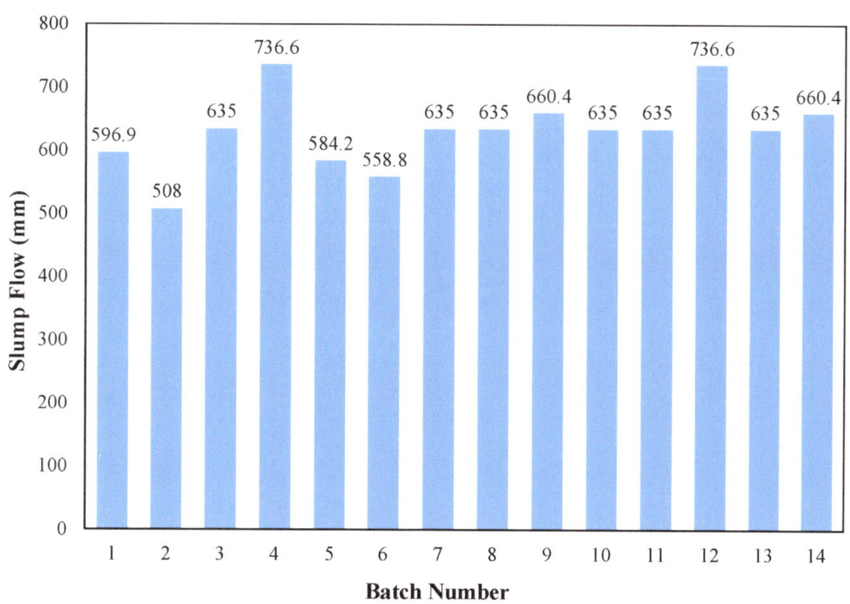

Figure 1.
Slump flows of SCFRC batches.

was recorded for batch 3. This air-content values were favorable, as compared to the results reported elsewhere [30]. The experimental results from batch 14 showed an excellent slump flow of 660.4 mm with an air content of 4.8 percent.

4.3 Compressive strength

The compression tests were carried out for the SCFRC cylinders to obtain the three-day and seven-day compressive strengths. **Figure 3** shows the compressive

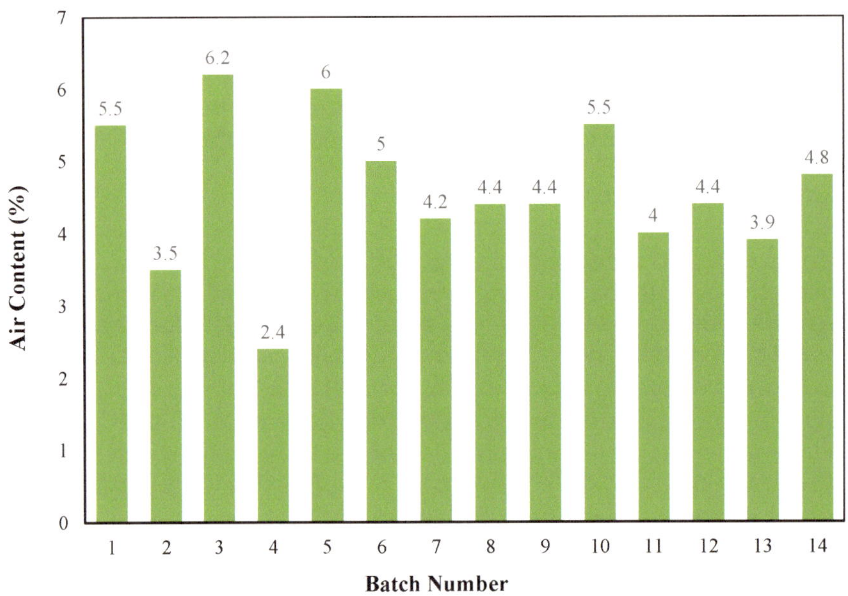

Figure 2.
Air contents of SCFRC batches.

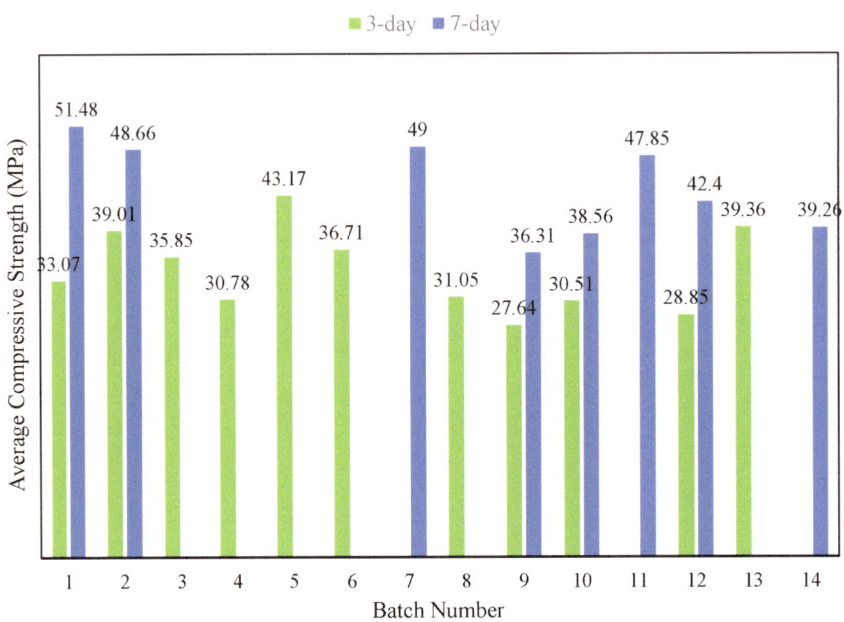

Figure 3.
Average compressive strengths of SCFRC batches.

test results of the SCFRC specimens from selected batches at three and seven days. In some batches, due to the premature failure of some cylinders, the results at the three- and seven-day compressive testing were not reported in the figure. The compressive strengths ranged from 27.64 and 43.17 MPa for three-day testing specimens while the seven-day testing results showed compressive strengths ranging from 36.31 to 51.48 MPa, which well fulfilled the application needs.

The average three-day compressive strength of batch 3 (35.85 MPa) was lower than that of batch 2 (39.01 MPa). The reduction in the three-day compressive strength of cylinders in batch 3 was due to the increase of water and the coarse aggregate adjustments. The SCFRC strength of the three-day compressive test results of batch 5 (43.17 MPa) exhibited the highest compressive strength among all the tested batches. Several trials were conducted after this batch, and the desired mixture was attained in batch 14 with the seven-day SCFRC compressive strength of 39.26 MPa.

Figure 4 shows the compression failure of selected cylinders from four batches. As can be seen in **Figure 4a**, the cylinder broke along the edge, not through the center. The failure is attributed to a minor defect in the bottom surface of the cylinder molds. The mold irregularity effect on the compressive strength were taken care of while producing subsequent batches, as seen in **Figure 4b**. The cylinders in batch 3 failed at the middle section and concrete splitting at the bottom of the cylinder was observed.

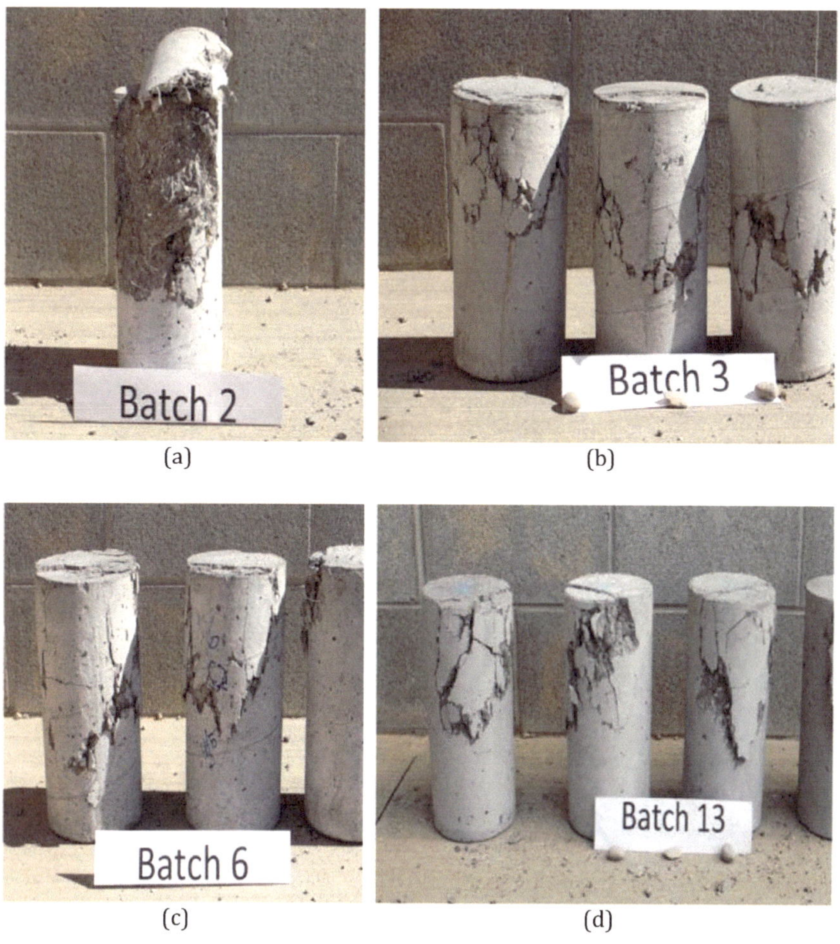

Figure 4.
Compression failure of SCFRC specimens from four batches (a) batch 2, (b) batch 3, (c) batch 6, (d) batch 13.

4.4 Esthetic characteristics

Production of esthetically pleasing SCFRC mixtures was a major objective of this study. For each SCFRC batch, several samples were cast to evaluate the esthetic characteristics of SCFRC. **Figure 5** shows a desired surface appearance of the SCFRC samples made from the selected batches. Overall, the color variation from a specimen to another was negligible. The specimens were easily removed from the molds, but the quality of the artworks in terms of sharpness and details varied significantly among the concrete batches.

As can be seen in the SCFRC specimen from batch 5 (**Figure 5a**), the artwork was clear but some minor surface voids were present. The quality of the artwork was significantly improved after several trials. Slight air voids on the surface of concrete from batch 7 were observed (**Figure 5b**). Excellent results were obtained for the concrete shapes made from the SCFRC mixtures in batch 9. As shown in **Figure 5c**, the artwork had almost no surface voids and a smooth finish. **Figure 5d**

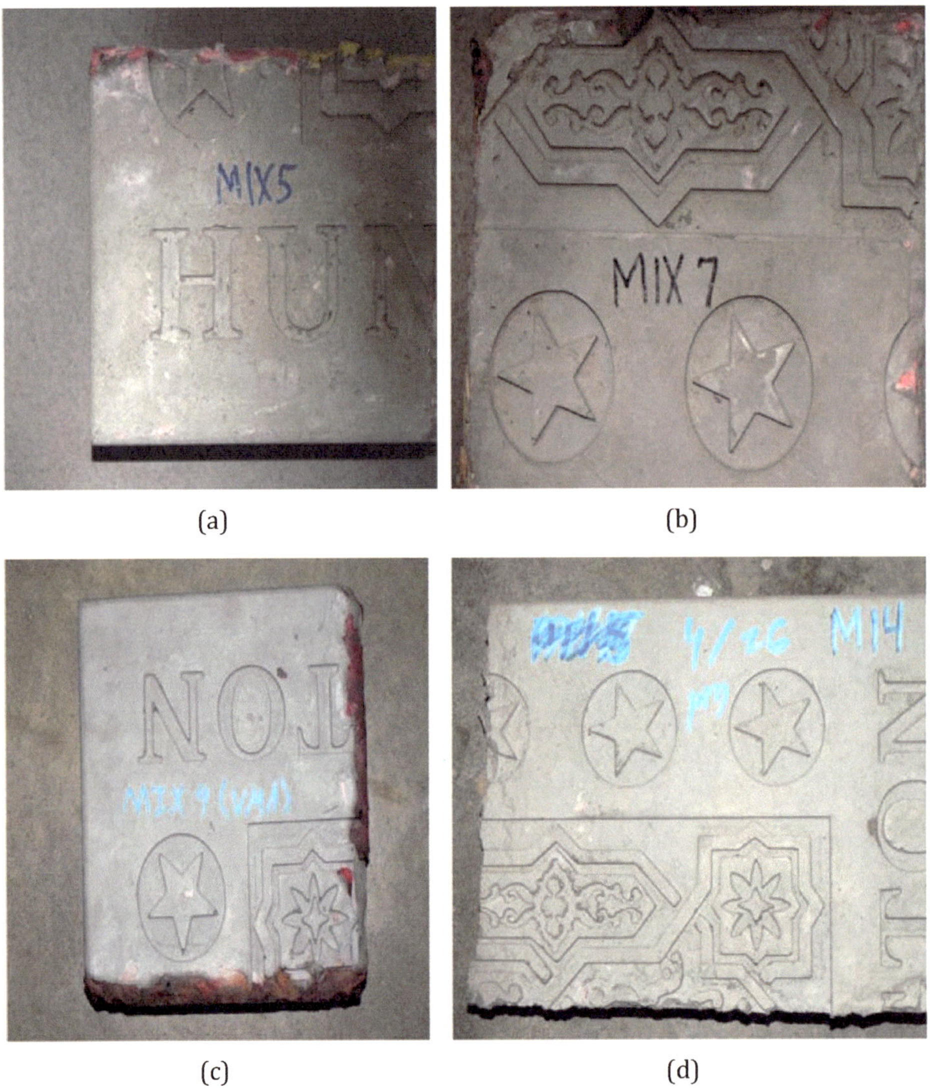

Figure 5.
Smooth finish appearance of SCFRC samples made from different batches: (a) batch 5, (b) batch 7, (c) batch 9, (d) batch 14.

exhibited the esthetic results of the specimen from batch 14. The specimen looked flaw free with a smooth finish and virtually no surface voids.

5. Conclusions and future works

Several mix designs of self-consolidating fiber-reinforced concrete (SCFRC) were experimentally examined. Properties of SCFRC mixtures made from micro silica, fly ash, and PP fibers were evaluated. The trial-and-adjustment method was employed to seek finding favorable performances of the SCFRC mixtures. The fresh concrete properties, including filling ability and air content, were evaluated. In addition, both the hardened concrete properties and esthetic characteristics of the mixture designs were examined. The trial-and-adjustment process produced SCFRC mixtures with desired material properties and esthetic appearance. The seven-day compressive strength of SCFRC ranged from 36.31 MPa and 51.48 MPa, and the average air content was approximately 4.5 percent. The developed mixture designs in this study were proven to be advantageous for potential SCFRC applications in architectural structures including building façades and esthetically-pleasing bridges.

Future work will focus on building a comprehensive database for desired SCFRC mixtures taking into account additional variables including fiber type (natural fibers, synthetic fibers, or the combination thereof) and chemical admixtures (e.g., plasticizers and superplasticizers). Such a study could assist with mass application of SCFRC in sustainable architectural structures.

Acknowledgements

The authors would like to acknowledge the support provided by all the members of the Concrete Materials Laboratory at Marshall University. Special acknowledgement is due to Dr. Tu Nguyen for his earlier contribution to drafting text that summarizes the research records compiled by the research team.

Conflict of interest

The authors hereby declare no potential conflicts of interest with respect to the research, authorship, and/or publication of this article.

Abbreviations

ACI	American Concrete Institute
AEA	Air-entraining admixture
ASTM	American Society for Testing and Materials
C-SCC	Colored self-consolidating concrete
FRC	Fiber-reinforced concrete
GBFS	Granulated blast furnace slag
HRWR	High-range water-reducing admixture
PP	Polypropylene
SCC	Self-consolidating concrete
SCFRC	Self-consolidating fiber-reinforced concrete
VMA	Viscosity-modifying admixture
VSI	Visual stability index

Author details

Wael Zatar* and Hai Nguyen
College of Engineering and Computer Sciences, Marshall University, One John Marshall Drive, Huntington, WV, USA

*Address all correspondence to: zatar@marshall.edu

References

[1] Khayat KH. Workability, testing, and performance of self-consolidating concrete. Materials Journal. 1999 May 1;96(3):346-53. DOI: 10.14359/632.

[2] Okamura H, Ozawa K. Mix design for self-compacting concrete. Concrete library of JSCE. 1995 Jun 25;25(6):107-20.

[3] Okamura H, Ouchi M. Self-compacting high performance concrete. Progress in Structural Engineering and Materials. 1998 Jul;1(4):378-83. DOI: http://dx.doi.org/10.1002/pse.2260010406.

[4] Okamura H, Ozawa K, Ouchi M. Self-compacting concrete. Structural concrete. 2000 Mar;1(1):3-17. DOI: http://dx.doi.org/10.1680/stco.2000.1.1.3.

[5] Okamura H, Ouchi M. Self-compacting concrete. Journal of advanced concrete technology. 2003;1(1):5-15. DOI: https://doi.org/10.3151/jact.1.5.

[6] Ouchi M, Nakamura SA, Osterberg T, Hallberg S, Lwin M. Applications of self-compacting concrete in Japan, Europe and the United States. Kochi University of Technology, Kochi, Japan. 2003 Oct 20.

[7] Goodier CI. Development of self-compacting concrete. Proceedings of the Institution of Civil Engineers-Structures and Buildings. 2003 Nov;156(4):405-14. DOI: https://doi.org/10.1680/stbu.2003.156.4.405.

[8] Domone PL. Self-compacting concrete: An analysis of 11 years of case studies. Cement and concrete composites. 2006 Feb 1;28(2):197-208. DOI: https://doi.org/10.1016/j.cemconcomp.2005.10.003.

[9] Sahmaran M, Yurtseven A, Yaman IO. Workability of hybrid fiber reinforced self-compacting concrete. Building and Environment. 2005 Dec 1;40(12):1672-7. DOI: https://doi.org/10.1016/j.buildenv.2004.12.014.

[10] Zollo RF. Fiber-reinforced concrete: an overview after 30 years of development. Cement and concrete composites. 1997 Jan 1;19(2):107-22. DOI: https://doi.org/10.1016/S0958-9465(96)00046-7.

[11] Lura P, Terrasi GP. Reduction of fire spalling in high-performance concrete by means of superabsorbent polymers and polypropylene fibers: Small scale fire tests of carbon fiber reinforced plastic-prestressed self-compacting concrete. Cement and Concrete Composites. 2014 May 1;49:36-42. DOI: https://doi.org/10.1016/j.cemconcomp.2014.02.001.

[12] Sonebi M, Grünewald S, Cevik A, Walraven J. Modelling fresh properties of self-compacting concrete using neural network technique. Computers and Concrete. 2016 Oct 1;18(4):903-20.

[13] ACI 237R-07, Self-consolidating concrete. American Concrete Institute, Farmington Hills, MI, 2007.

[14] Bartos PJ, Sonebi M, Tamimi AK, editors. Report 24: workability and rheology of fresh concrete: compendium of tests–report of RILEM Technical Committee TC 145-WSM. RILEM publications; 2002.

[15] Daczko JA, Kurtz MA. Development of high volume coarse aggregate self-compacting concrete. InProceedings of the Second International Symposium on Self-Compacting Concrete, Tokyo, Japan 2001 Oct 23 (pp. 403-412).

[16] Su N, Hsu KC, Chai HW. A simple mix design method for self-compacting concrete. Cement and concrete research. 2001 Dec 1;31(12):1799-807.

DOI: https://doi.org/10.1016/S0008-8846(01)00566-X.

[17] Sonebi M, Grünewald S, Walraven J. Filling ability and passing ability of self-consolidating concrete. ACI Materials Journal. 2007 Mar 1;104(2):162.

[18] Felekoğlu B, Türkel S, Baradan B. Effect of water/cement ratio on the fresh and hardened properties of self-compacting concrete. Building and Environment. 2007 Apr 1;42(4):1795-802. DOI: https://doi.org/10.1016/j.buildenv.2006.01.012.

[19] Uysal M, Sumer M. Performance of self-compacting concrete containing different mineral admixtures. Construction and Building materials. 2011 Nov 1;25(11):4112-20. DOI: https://doi.org/10.1016/j.conbuildmat.2011.04.032.

[20] López A, Tobes JM, Giaccio G, Zerbino R. Advantages of mortar-based design for coloured self-compacting concrete. Cement and Concrete Composites. 2009 Nov 1;31(10):754-61. DOI: https://doi.org/10.1016/j.cemconcomp.2009.07.005.

[21] Corinaldesi V, Monosi S, Ruello ML. Influence of inorganic pigments' addition on the performance of coloured SCC. Construction and building materials. 2012 May 1;30:289-93. DOI: https://doi.org/10.1016/j.conbuildmat.2011.12.037.

[22] Ghalehnovi M, Shamsabadi EA, Khodabakhshian A, Sourmeh F, de Brito J. Self-compacting architectural concrete production using red mud. Construction and Building Materials. 2019 Nov 30;226:418-27. DOI: https://doi.org/10.1016/j.conbuildmat.2019.07.248.

[23] ASTM C150/C150M-18, "Standard Specification for Portland Cement." ASTM International, West Conshohocken, PA, 2018.

[24] ASTM C33/C33M-18. "Standard Specification for Concrete Aggregates." ASTM International, West Conshohocken, PA, 2018.

[25] ASTM C618-19. "Standard Specification for Coal Fly Ash and Raw or Calcined Natural Pozzolan for Use in Concrete." ASTM International, West Conshohocken, PA, 2019.

[26] ASTM C1017/C1017M-13e1. "Standard Specification for Chemical Admixtures for Use in Producing Flowing Concrete." ASTM International, West Conshohocken, PA, 2013.

[27] ASTM C260/C260M-10a. "Standard Specification for Air-Entraining Admixtures for Concrete." ASTM International, West Conshohocken, PA, 2016.

[28] Zatar, Wael, and Tu Nguyen. "Mixture Design Study of Fiber-Reinforced Self-Compacting Concrete for Prefabricated Street Light Post Structures." Advances in Civil Engineering 2020 (2020).

[29] S. Nagataki and H. Fujiwara. "Self-Compacting Property of Highly-Flowable Concrete." ACI Special Publication SP-154, pp. 301-314, 1995.

[30] Barfield, M., & Ghafoori, N. "Air-entrained self-consolidating concrete: A study of admixture sources." Construction and Building Materials, 26(1), 490-496. 2012. https://doi.org/10.1016/ j. conbuildmat. 2011.06.049.

Section 3

Urban Development

Chapter 7

Extensive African Urbanization: The Case of the Mozambican Periphery

Joaquim Miranda Maloa

Abstract

The aim of this article is to analyze African urbanization, looking specifically at the transformations of Mozambican urban peripheries. To do so, conduct a qualitative research of an exploratory nature, using bibliographic, documentary and photographic survey. The most interesting discoveries of this study started in the 1990s, a period marked by the end of "socialism", civil war, centrally planned economy, which verified the opening of parents in the current Western capitalist world in the growth of foreign investments and financial transactions. Under the effect of market liberalization. At this moment, everything that is traded and the exchange value overlaps the use value and appropriation of space in exchange for money. In this context, as the peripheral urban cities are transformed into multiplying duplexes, elegant houses those call houses, true "Mozambican palaces" and closed condoms. The establishment of these houses will transform these spaces and gradually expel the low-income population that has existed for a long time, to places very far from the central area, this phenomenon is called gentrification. He concluded that the transformation of Mozambican urban peripheries is influenced by the increase in real estate capital, increase in individual income, ease of acquisition of space and construction material (provided by the expansion of the installation for the exhibition that makes it possible or cheaper). This research is important because it makes an important contribution to the empirical studies on the new neoliberal urbanism that is taking place in Mozambican cities. The limits of this research are due to the lack of funding to carry out a systematic survey of new ventures that will emerge in cities and places far from the center of large cities, such as: *Maputo*, *Beira, Nampula* and *Matola*. It is intended in the future study to demonstrate how to change the socioeconomic structure of the residents of the Mozambican peripheries, characterizing gentrification.

Keywords: extensive urbanization, transformation, urban spaces, periphery, gentrification

1. Introduction

Three geographic objects, among many others, can illustrate the transformations of the built environment of the Mozambican urban peripheries: the duplexes - high-end properties composed of two floors interconnected by stairs [1]; elegant houses, which we call houses, real "Mozambican palaces" and closed condominiums [2]. These urban transformations are influenced by the increase in individual income, ease of acquisition of space and construction material (provided by the expansion of the establishment for acquisition that makes it cheaper), which change the existing residential landscape and restructure the peripheries, valuing, making it more attractive and promoting its insertion in the urban market, causing negative impacts related to processes of gentrification and socio-spatial segregation [3].

To use the words of Diniz and Silva [3], the ennoblement of the urban periphery occurs through two social substitution processes, linked to urban rehabilitation actions through private investments and culminating in the expulsion of the former residents. The first that goes from 1990 to 2000, when the low-income population left the central areas of the main cities, to the peripheries. These low-income residents sold the keys (a typical Mozambican expression meaning the sale of an apartment or flat), while others rented their houses, with or without contracts, and went to live on the outskirts. In a second moment, since the beginning of the 2000s, the formation and consolidation of the modern real estate market and the civil construction industry accelerate the pace of construction of new housing and commercial areas in the peripheries [4].

It is in this context that a new gentrification arrives in Mozambican cities, first in urban centers, later in urban peripheries, connecting distant points that are points of infrastructures such as highways, which value an appropriation of space.

From this perspective, it is based on the principle that the poor population is the greatest vulnerability in these spaces, in each moment when new ventures are implemented, they are motivated to leave the neighborhood spaces because they verify that a new life does not correspond to their financial capacity, considering as profound social inequalities in urban peripheries, it is possible to anticipate the geographic assessment of changes in the socioeconomic structure of residents of the Mozambican peripheries, characterizing gentrification. In this way, this work starts from some important issues in relation to the transformations of the peripheries of Mozambican cities, but it can also be replicated in other African cities. The first specific objective is to map the types of housing and commercial developments that are emerging in the urban peripheries. The second is to present a spatial distribution of the suburbs in urban cities.

In this way, therefore, this article is divided into three parts: in addition to this introduction, a methodology, with a description of the procedures used for writing that article and analyzing the results.

2. Methodology

To this end, a qualitative research of an exploratory nature was carried out, using bibliographical, documentary and photographic survey to show the dynamics of the transformations of the Mozambican urban peripheries. The hefty data were obtained between, 2012–2020. The researcher had the opportunity to visit the peripheries of the various Mozambican cities, such as: *Maputo, Matola, Xai-Xai, Maxixe, Beira, Chimoio, Tete, Beira, Nampula, Nacala, Lichinga* and other small cities, with a capital, with a view to verifying the changes that may occur in the peripheries, for the closer approximation of the socio-spatial reality that is being investigated. The data were collected in each city through informants who took us to each peripheral neighborhood, arriving there and contacting the secretaries of the neighborhoods that accompanied us in each transforming block and asking for authorization of the requirements to take as photographs of the high and elegant of the neighborhoods. Condominiums. The data were collected through photographs, these data were obtained through a comparison between the peripheries of the cities to reach the conclusion of the transformations of the urban peripheries. These transformations are perceived by the emergence of new elegant buildings and condominiums, which are pushing the poorest people to places further away from these peripheries.

The precision of the data led us to understand the urban gentrification index, through two stages: (i) The selection of variables and definition of indicators, such as the type of housing existing before or after the dimensions; (ii) standardization and weighting of urban indicators, where urban housing types were compared.

For the accuracy of the data, we used two stages: The first stage referred to the search for documents and a consistent bibliography regarding urban studies in Mozambique, which helped in the identification of social conditions that literature in the field of Geography presents as consolidated socio-spatial realities no urban territory in the city. Thus, among a series of works, the study by Muacuveia [5], which evaluated the residential segregation of the city, as well as the work of Maloa [6], in the analysis of exclusion in the urban area, is worth mentioning. Literature on urban dispersion was also observed, with authors such as Maloa and Nascimento [7]. It is also worth mentioning that these literatures directly infer on an extensive urbanization. The second stage of the research sought to standardize the evidence, with some tables, graphs, graphics or visual materials such as photographs from the internet to improve the quality of the article's communication.

3. Results

3.1 The first phase of extensive urbanization the rise of duplexes and elegant houses in the suburbs

The first phase of the process of fostering the urban periphery, here Gutiérrez [8] calls it gentrification, was a phenomenon studied by British society Ruth Glass in his book "London: aspects of change" [9]. The term gentrification is derived from the English noun "gentry", which designates "successful" individuals or groups. The term was created to refer to a process of elitization or "enrichment" of local places in the city, previously characterized as predominantly popular areas. The same author recognizes the class character of social inequalities added to this phenomenon [10].

However, it was the Scottish geographer Neil Smith (1954–2012), based in the United States of America, who analyzed in depth the various processes of promoting poor neighborhoods in New York, with emphasis on the Harlem neighborhoods in Manhattan, one of New York [11].

Gentrification is an urban and social phenomenon, altered by the development of the degraded housing stock of the popular classes through its gradual investment by the middle class [12]. It is also included, as a process of urban restructuring, marked by the economic restructuring characteristic of "late capitalism" and advanced, conditioned by a more flexible subsidiary capital accumulation regime [13].

Since the mid-1990s, as main Mozambican cities such as: Maputo, Matola, Beira, Inhambane, Xai-Xai, Beira, Chimoio, Tete, Quelimane, Nampula, Pemba and Lichinga, are undergoing urban changes that end up renovating or together geographical, social and political relations producing new urbanities [14] and establishing security standards very different from previous periods [15].

At the time, as transformations in the urban peripheries were driven by the country's opening up to the western capitalist world, by the growth of foreign investments and financial transactions under the effect of market liberalization, global dynamics that transform urban spatiality.

The restructuring of the urban space gives rise to new ventures, in addition to the old periphery [16], and emerging functionalities, with the entry of new urban actors, these strategies are influenced by the furniture market and the willingness of residents to build and build a house own.

The first phase of the transformation of peripheral urban spaces has historically been influenced by three movements that, although contradictory, refer to one another and interpenetrate. The first movement concerns the arrival, in the early 1990s, of numerous international non-governmental organizations (NGOs) who came to support Mozambique who have just emerged from a protracted civil war. Its members started to rent apartments and houses, for housing or for transformation into apartments. Payment was made in US dollars. The result that the urban fabric of the central nucleus started to be affected by dollarization. Shopping streets, parks, restaurants, markets, museums, cinemas and all types of residential and commercial properties were valued by dollarization.

We can say that in the 1990s, or at the beginning of the "commercialization of the city" [17], marked by the flow of technicians from different western countries, it generated a speculation harmful to the population of the central area, with emphasis on the numerous civil servants installed there. Many of them then had to rent their houses.

The second movement was fueled by the "neoliberal spirit", which was strengthened by the end of the civil war, which allowed many Mozambican citizens to start to benefit from the rental income from properties located in central areas. Until this moment, the production of urban life was guided by the survival strategy.

And the third space–time corresponded to the end of the nationalization of the buildings, protected by Decree-Law 5/76. For its part, Decree-Law 2/91 of January 1991, instituted the sale of properties previously nationalized by the State [18]. Article 1 authorizes Mozambicans in a legal situation whose properties are supervised by the Administration of the State Real Estate Park (APIE), to sell, rent, restore, etc. the same, as long as he asked the municipality for authorization [19].

The alienation law increased the profitability of real estate in the main Mozambican cities. High inflation motivated homeowners to rent their properties as a survival mechanism. We can, then, declare that, in the 1990s, it was the decade of organizing the rental of houses, which allowed the fringe of the urban population affected by inflation above the rate of increase in economies, to escape bankruptcy. The "low-income" residents [20] who acquired real estate during the nationalization period from "peripheral neighborhoods", looking for places where they could build new houses to live and survive with income in the city. The rent of houses becomes a source of family income, especially for those who had waste during the nationalization of buildings in 1976.

Many Mozambicans who were able to buy APIE real estate at the time of the sale in the 1990s, and sell or sell to live on the outskirts, contributed to financial and real estate inflation. The new residents build houses with cement and zinc blocks, while losing space like reed houses that exist [16]. Peripheral spaces are used more hybrids. Elegant houses coexist, reed, cement, zinc sheet, cardboard etc. [21]. In the process of this process, imposing houses and elegant duplex type between conventional dwellings that replaced the precarious ones (**Figures 1, 2** and **3**)[1].

Duplexes and elegant houses are spatial manifestations of globalization, which are reproduced from the process of constituting the new urban society, supported by the expansion of the world market, the selection of borders and the commodification of space. As goods, their main characteristics are produced by private builders [23] influencing planetary styles of real estate consumption. These buildings, driven by the desire for home ownership, a central element of consumption in the contemporary world, transform like urban landscapes of the peripheries,

[1] To see: https://www.archdaily.com.br/br/country/mocambique.

Figure 1.
A duplex residence on the outskirts of Maputo. Source: [22].

Figure 2.
A duplex residence with precarious houses on the outskirts of the city of Lichinga. Source: [22].

as they gradually appear in precarious areas. It is incorrect to say that the idea of home ownership reached only the "middle class" population or individuals with average purchasing power. Many high-ranking government officials searched the new suburbs for spaces to build their dream homes.

3.2 Second phase of extensive urbanization: the emergence of condoms in peripheral neighborhoods

The private real estate sector with national and international capital started to gain relevance in Mozambique mainly in the 2000s, imposing a new financing model within the constitution of a commercial financial system[2]. As we know, a post-colonial Mozambican urbanization had little accommodative solution to the

[2] To see: https: //viveremmocambique2017.blogspot.com/2018/07/morar-em-mocambique-tete_23.html.

Figure 3.
A duplex residence on the outskirts of the city of Lichinga. Source: [22].

Figure 4.
*Transformation of the peripheral urban spaces of the city of Tete (*Vale dos Embondeiros Condominium with its wide streets and its various Embondeiros*). Source: [24].*

housing problem of senior state officials and a group with rising purchasing power[3]. These have been used since the 2000s, symbols of consumption of the real estate machine. According to **Figures 4, 5** and **6**[4].

In the 2000s, many civil construction companies, homebuilders, appeared at a time when urbanization was intensifying and the job market with high wages was expanding while many foreign professionals and large corporations entered Mozambique [27]. Ten years after the end of the civil war, the country was becoming safe to attract workers and international capital. In this phase, the transition

[3] To see: http://hikersbay.com/africa/mozambique/hotel/mz/condominio-veloso.html?lang=pt.

[4] To see:https://www.jumia.co.mz/arrenda-se-moradia-t4-no-condominio-khurula-pid6726355.

Figure 5.
Transformation of the peripheral urban spaces of the city of Nacala. [25].

Figure 6.
Transformation of the peripheral urban spaces of the city of Maputo, Bairro T4, in the Khurula condominium. Source: [26].

from the rentier property model in the urban nucleus to the renterist real estate promotion model, aimed at condominiums in the peripheries, which generate gentrification and dispersion of high-end condominiums and houses (duplexes), which cohabit, takes place. With irregular occupations [28]. The new architectural landscape is influenced by forces of globalization that produce a hegemonic rationality through the pragmatic discourse of the civil construction industry sector in the real estate market, in order to meet the needs of the restricted social segment, gathering the necessary conditions for indebtedness with institutions [29, 30].

The foreign capital flows that enter the real estate circuit of Mozambique, generate questions for social interests and critical thoughts for studies on urbanism and financialization, at a time when the right to urban infrastructure and services faces new challenges with the increase of selected private investment, marked as conditional on real estate financing, such as: a regular salary or income, a low level of performance, with the financing charges; payment of life insurance [19];

among other requirements to optimize capital accumulation and the difficulties of the peripheries in response to universal requirements for urban rights [31].

Law governing the construction of condominiums - Decree of 17/2013 [32], contributes to the increase in the production of real estate, while expanding real estate speculation and its effects, with a stock of land or empty lots waiting to be transformed into goods, as a basis for creating fictitious capital [33]. To use as Marina Fix's words [31], there is a strategy for using empty spaces, a promise of gains in the anticipated future or that can be created on land purchased and imported as obstacles to access infrastructures and urban services of frequent use time [34]. Extent to which closed condominiums have the capacity to install their equipment such as: water pump, transformers and energy generators and other facilities [32], public investment in infrastructure and urban services are discouraged, excluding the needy population. That resides in limited spaces. Therefore, the expansion of condominiums and the luxury duplex is creating a movement of social and housing differentiation in the peripheries [35]. "Capital transforms the spaces it finds in the spaces of production and accumulation (...). It is the only way to produce, destroy and recreate new bases, spaces and conditions for its expansion, construction and expansion of markets and expanded horizons for valorization" [31].

Among the residents who live in high-end condominiums are foreigners of various nationalities. Condominium flows are associated with the need for international capital, if they are captured and accumulated in the real estate circuit [36].

Currently, in urban Mozambique, most luxury residences are located in peripheral neighborhoods, privileged spaces for gentrification and metamorphosis of the landscape that was once dominant, in the caniço [19, 37].

Accelerated urban periphery results partly from the presence of empty spaces, low prices on the ground and irregular development and flexibility in the urban land market that integrates the process of capital accumulation by real estate [13].

A "link between the land market and the capital market is a resource of financial capitalism" [31], p. 3. The gentrification process results in part from the accumulation of capital by real estate agents that meet the specific purpose of the gentifiers ("middle class" and "upper middle class", who are filtering them out - relatively wealthy) at the expense of broad social interest [38].

Final considerations.

As pointed out, in the years before 1990 as peripheries they were places devalued by individuals with high purchasing power [4]. Since then, peripheral spaces have expanded as places covered by these citizens and attract service ventures - stores, banks, ATMs and small businesses, benefits for the interests of new residents that, in turn, justify the growth of public investments in these areas. You can also check in several cities or at florists of equipment such as bus terminals or shopping centers and specific activities that exist only in the central area. Meanwhile, the value of urban land increases with new condominium developments.

Therefore, a real estate construction industry generates in the peripheries a speculation of land that motivates the old residents to sell their land (lots). Therefore, according to Maria Monica Arroyo, as urban peripheries they are configurating mosaics of multiple combinations of temporalities, diversities, opposition, showing many ways of doing, feeling and living [39].

The second phase of the transformation of peripheral urban spaces is being accompanied by the growing role of the private circuit of private capital (national and international), driven by real estate developers and developers interested only in profits. For Atkison and Bridge [40], this reality is the starting point of a new urban colonialism. The increase in habitual high-standard demand in the peripheries, which creates fragmentation and disconnection, is a reflection of the lack of a national housing policy to understand all classes, but also an unremitting pursuit of profits through the commercial market. The process confirms that urban areas are increasingly attuned to the pace of the globalized world [39].

What is currently happening in Mozambique is a true capitalist production in the cities [41], achieved by the self-financing of those with a high purchasing power, especially for those with variable income between 8 and 15 years old, and can use part of monthly resources in a house in the condominium. This combination between the real estate industry and the neoliberal economy transforms as peripheries and, at the same time, food and gentrification, changes such as social inequalities, urban poverty and urban segregation. This shows an expansion of a real estate market with harmful consequences for a low-income population.

Author details

Joaquim Miranda Maloa
Department of Geosciences at Rovuma University - Niassa Extension, Lichinga, Mozambique

*Address all correspondence to: joaquimmaloa@gmail.com

References

[1] AGMAR. *As vantagens dos apartamentos duplex.* Disponível em: http://blog.agmar.com.br/as-vantagens-dos-apartamentos-duplex/. Acessado em 13.06.2020.

[2] CASTEL-BRANCO, Nuno. Opções Econômicas de Moçambique 1975-95: Problemas, Lições e Ideias Alternativas. In: BRAZÃO, Mazula (editor). *Moçambique Eleições, Democracia e Desenvolvimento*. Maputo: Elo gráfica, p. 581-636, 1995.

[3] Diniz, L. S., & Silva, C. P. B. Reestruturação estratégica e gentrificação em Belo Horizonte: novo cenário, velhas medidas. *Urbe. Revista Brasileira de Gestão Urbana*, 2019, p.1-15.

[4] SILVA Arménio. *Dinâmica soco espacial e produção habitacional na periferia de Maputo-Moçambique a partir da década de 1970:* destaque para bairros Polana Caniço "A" E "B". Dissertação (Mestrado em Geografia) - Universidade Federal de Santa Catarina. Santa Catarina, 2011.

[5] MUACUVEIA, Reginaldo. *Urbanização contemporânea em Moçambique: o papel dos instrumentos de planejamento urbano na ocupação do espaço.* Tese de Doutorado. Instituto de Geografia. Universidade Federal de Uberlândia.2019.

[6] MALOA, J. M. *A Urbanização Moçambicana Contemporânea: suas Características, Dimensão e Desafios*. Revista URBE. Revista Brasileira de Gestão Urbana (*Brazilian Journal of Urban Management*). 2019, p.1-15.

[7] MALOA, J. M.; NASCIMENTO Jr, Lindberg. *A Dispersão Urbana em Moçambique: Uma Contribuição ao Estudo de Produção do Espaço Urbano em Maputo*. RA'EGA – O Espaço Geográfico em Análise, Curitiba, Universidade Federal de Paraná, v.45 n.1, 2018, p.191-109.

[8] GUTIÉRREZ Rafael. La polissemia y la linguística de gentrificación. *Cadernos Metrópole,* v. 16, n. 32, p. 329-340, nov 2014.

[9] GLASS Ruth. Introduction. In: *London:* aspects of change. Centre for Urban Studies (dir.). Londres, Macgibbon & Kee, 1964, p. 12-41.

[10] PEREIRA Alvaro. A gentrificação e a hipótese do diferencial de renda: limites explicativas e diálogos possíveis. *Cadernos Metrópole* v. 16, n. 32, p. 307-328, nov 2014.

[11] SMITH, Neil. The gentrification of Harlem? *Annals of the Association of American Geographers*, v. 76, n. 3, 1986, p. 347-365.

[12] BENIS, Khadija. *Viela de Afama/ entre revitalização e gentrificação. Impacto da "Gentrificação." sobre apropriação do espaço público*. Dissertação de Mestrado em Estudos Urbanos e Mediterraneos- EURMed da Faculdade de Arquitetura da Universidade Técnica de Lisboa. Lisboa, 2001.

[13] MENDES Luis. Gentrificação e políticas de reabilitação em Portugal: uma análise crítica a luz da tese rent gap de Neil Smith. *Cadernos Metrópole*, São Paulo, v.16 n°32, 2014, p.487-511.

[14] SALVADOR, Cristina & RAPOSO, Isabel. Há diferença: ali é cidade aqui é suburbia, urbanidade dos bairros, tipos e estratégias de habitação em Luanda e Maputo. In: OPPENHEIMER, Jochen & RAPOSO, Isabel. *Subúrbio de Luanda e Maputo*. Lisboa: Colibri e Centro de Estudo sobre África e do Desenvolvimento, 2007.

[15] RAPOSO Isabel; SALVADOR Cristina. Há diferença: ali é cidade, aqui é subúrbio: urbanidade dos bairros, tipos e estratégias de habitação em Luanda e Maputo. In: OPPENHEIMER, Jochen & RAPOSO, Isabel. *Subúrbios de Luanda e Maputo.* Lisboa: Colibri e Centro de Estudos sobre África e do Desenvolvimento (Coleção tempo e espaços africanos). 2007, p.105-138.

[16] MENDES, Maria Clara. *Maputo antes da independência*: geografia de uma cidade colonial. Tese de Doutorado. (Geografia urbana) – Universidade de Lisboa, 1979.

[17] FERREIRA Álvaro. A cidade no século XXI: segregação e banalização do espaço. Rio de Janeiro: Consequência, 2011.

[18] CONDOMÍNIO JULIUS ESTATE. Disponível em: <juliusestate.com/assets/portolio-for-web.pdf>. Acesso em: 15 nov. 2015.

[19] MELO, Vanessa.; JORGE, Silva. Processos e dinâmicas de (re)produção do espaço (perí)-urbano: o caso de Maputo. SEGUNDO CONGRESSO INTERNACIONAL DE JOVENS INVESTIGADORES URBANOS (SICYURB). Lisboa, 11 a 14 de outubro de 2011.

[20] MUHAMBE Boavida. *A dinâmica contemporânea dos fluxos populacionais cidade/campo em Moçambique: o exemplo de alguns bairros do distrito da Matola*. Dissertação (Mestrado em Desenvolvimento e Cooperação Internacional) - Instituto Superior de Economia e Gestão da Universidade Técnica de Lisboa. Lisboa, 2000.

[21] SERRA Carlos. *Em cima de uma lamina*: um estudo sobre precariedade social em três cidades de Moçambique. Maputo: Imprensa Universitária-Universidade Eduardo Mondlane, 2003.

[22] MOTTELSON, Johan. *Residências compactas nos assentamentos informais*. 2020

[23] ASSOCIAÇÃO DE COMÉRCIO E INDÚSTRIA. *O quadro legal para obras de construção em Moçambique*. Maputo: GIZ, 2008.

[24] MOÇAMBIQUE. *Relatorio nacional da 3ª Conferência das Nações Unidas sobre Habitação e desenvolvimento urbano sustentavel – Habitat III*. Maputo: Ministerio das Obras, Habitacao e Recuros Hidricos, 2018.

[25] HIKERBAY. *Condomínio Santa cruz*. Maputo.2020.

[26] JUMIA, Mz. *Empresa imobiliária*. Maputo.2020

[27] TOPALOV Christian. *Les promotuers immobiliers*: *contribution à l'nalyse de la production capitaliste du logement em France*. Paris:Mouton e Co, 1974.

[28] DECRETO - LEI n° 5/76, *que regulamentava a nacionalização de prédios*. Boletim da República.

[29] COSTA Bernardo; HENRIQUE Cristina. Dinâmicas urbanas. Maputo e condomínios fechados, transformações espaciais e processos de mobilidade residencial. Sociedade de Geografia de Lisboa. Lisboa, junho de 2014. *Cadernos Metrópole.* v. 16, n. 32, p. 565-586, nov 2014.

[30] SOMEKH, Nadia. *A cidade vertical e o urbanismo modernizador, 1920-1939*. São Paulo: EDUSP, 1997.

[31] FIX, Marina. *Financerização e transformação recente no circuito imobiliario no Brasil.* Tese (Doutorado em Desenvolvimento Económico) - Faculdade de Económia da Universidade Estadual de Campinas. 2011.

[32] DECRETO n° 17/2013. *Regulamento do regime jurídico do condomínio*. Boletim

da República de Moçambique, I Série-número 34, Sexta-feira, 26 de abril de 2013.

[33] DECRETO - LEI n.° 2/91, de 16 de Janeiro, *determina os procedimentos concernentes à alienação* de imóveis. Boletim da Republica.

[34] DECRETO n° 15/2010. *Regime de contratação de empreitada de obras públicas*. Aprovado pelo Conselho de Ministros no dia 24 de maio do mesmo ano. Publicado no Boletim da República, I Série, n° 20, suplento, de 24 de maio de 2010.

[35] DIPLOMA MINISTERIAL 49/2013. *Aprova fórmulas de revisão de preços de empreitadas de obras públicas*. N° 42 de 24.05.13, Boletim da República - I Serie, p. 33

[36] MALOA, J. M. *A construção jurídica do espaço urbano em Moçambique: uma reflexão a partir da norma jurídica*. CaderNAU – Cadernos do Núcleo de Análises Urbanas, v 10, n1, 2018, p.157-170.

[37] MENDES, Maria Clara,. FERNANDES, Mário. *Dicotomias urbanas em Moçambique*: cidades de cimento e de caniço.2012 [s.l..s.n] Disponível em < sigarra.up.pt/flup/en/publs_pesquisa> Acesso em: 2 jan. 2013.

[38] SMITH, Neil. Toward a theory of gentrification: a back to the city movement by capital not people. *Journal of the American Planning Association*, n. 45, 1979. p. 538-548.

[39] ARROYO, Mónica Maria. A economia invisível dos pequenos. *Le Monde Diplomatique Brasil*, V ano 2, 2008, p.30-31.

[40] ATKINSON, R. e BRIDGE, G. *Gentrification in a global context:* the new urban colonialism. Londres, Routledge, 2005.

[41] MARICATO, Ermínia. *A produção capitalista da casa (e da cidade) no Brasil industrial*. Editora Alfa-Omega, 1979.

Chapter 8

Traffic Flow Analysis and Management

Tsutomu Tsuboi

Abstract

This research is about joint government founded program between Japan and India or Science and Technology Research Partnership for Sustainable development (SATREPS). The purpose of this research is to establish Low Carbon Transportation in developing countries and we choose one of major city in India, where it is Ahmedabad city of Gujarat state—west cost of India. In order to approach the target, we need to understand the current situation of traffic condition in the city. The current traffic condition in India is some chaotic because of their different driving behavior compared with the advanced countries. It is becoming the chaotic traffic condition in India by not only diving behavior during investigation of this research. The main reason of the traffic congestion comes from the unbalance between growing transportation demand and its insufficient infrastructure preparation. In this chapter, it introduces the current traffic condition based on four years monitoring of the traffic by the traffic monitoring cameras and comparison by the traffic flow theory at first. Then it introduces the new traffic analysis method especially for its traffic congestion analysis and its parameters. After the traffic congestion analysis, it summarizes conclusion and our next step from the experience.

Keywords: traffic flow, intelligent transport system, traffic density, traffic volume, developing country transportation

1. Introduction

Transportation is important for human been activity as the economy grows. Under rapid growing economy, it always happens negative impact by growing transportation because of i-balance between transportation demand versus infrastructure development such as road construction, traffic control management system, public transportation implementation, and so on. Negative impact occurs as the following—traffic congestion, traffic accident, un-necessity fuel consumption, environment destruction, air prolusion, noise pollution, traffic fatality—as transportation issues in general. There is no exception bout this transportation issue in the world, not only developing countries but also advanced countries. Therefore it is important to understand traffic flow condition and mechanism of traffic congestion reason at least.

In order to understand traffic flow, it is necessary to measure the traffic condition by traffic monitoring system such as traffic video monitoring camera, other sensing technology for directly collection vehicle movement like smart phone, navigation system in vehicles, drone camera, 3D sensor camera, infrared camera, Bluetooth device in vehicle, and many other high-technology tool available. But the point is how to collect accurate traffic data as real traffic flow and under developing counters, it becomes challenge to understand the real traffic condition from the data rather than in advanced counties. In this book, it is introduced traffic flow analysis method combined with traffic flow theory.

After collecting traffic data and analysis, it digs into traffic flow problem and shows important traffic flow parameter for traffic congestion with using actual traffic measurement data in a major city in India. And it shows city level traffic flow visual analysis by Geographic Information System (GIS) tool to understand its traffic issues and some countermeasures.

Based on this project, it mentioned what we learn and how to approach for solving the issues. Technology becomes higher and provides us better solution. In conclusion, author wants to share the idea that technology is one of tool for helping collecting data and we need to know how to creating our experience and how to use it as sustainable concept. At last part, author adds traffic fatality issues from fifty years Japanese experience with some enhanced experimental equation in appendix.

2. Case study of project

2.1 Case study field

This project has been started since April 2017 under SATREPS program, which is government founded joint research project between Japan and India. The case study filed was chosen in the Ahmedabad city of Gujarat state India, which is one of typical rapid economic growing and faced to serious traffic issues. The population of Ahmedabad city is over 8 million [1] in 2018 from 5 million in 2011 and the number of vehicles is about 4 million in 2017 [2]. More than 70% vehicle is two wheelers, which is typical percentage in developing countries. Based on city government or Ahmedabad Municipal Corporation (AMC) agreement, the case study field was chosen in west side of the city called "New City" where there will have more commercial business and new building constructions (**Figure 1** and **Table 1**).

2.2 Monitoring traffic in Ahmedabad

First, it is necessary to have monitoring tools for measuring traffic condition in the city. In this program, we installed traffic video monitoring cameras shown in **Figure 2**. This is general camera in traffic industry. And the special high resolution 4 K camera is installed at the major junction—Pladi junction as detail analysis at junction.

Second, the total number of cameras is 36 including 4 K camera. The camera location is shown in **Figure 3** (The number is camera ID and 4 K camera has no number, but it is in the center among 2000s ID cameras in the map). The traffic video camera has several functions such as counting number of vehicles on the road, average vehicle speed, traffic density, occupancy and so on. Traffic density is the number of vehicles per kilo meter on the road (vehicle/km) which is defined in the traffic flow theory. Occupancy is how much percentage occupied by vehicles on the road which is also defined in the traffic flow theory. The detail parameter explanation is described in later of this chapter.

Third, here is an example of traffic condition data which is shown in **Figure 4**. There are two graphs shown which are time-based traffic volume and average vehicle speed. From those graphs. we see the traffic condition (in this case camera #2 location). It is clear that the traffic congestion was occurred around 20:00 o'clock because its traffic volume was peek and average vehicle speed is lowest. The data here is a month data in June 2019 and each graph shows average point and standard deviation.

Figure 1.
Ahmedabad city map [3].

Co-ordinates:	23.03° N 72.58° E
Area:	466 Sq.km. (year 2006)
Population:	55,77,940 (year 2011 Census)
Density:	11,948 /sq.km
Literacy Rate:	89.60%
Average Annual Rainfall:	782 mm
Popularly known as:	Amdavad
STD Code:	079

Table 1.
Profile: Ahmedabad City [3].

2.3 Traffic flow theory and measurement

2.3.1 Traffic flow theory

In this section, it is necessary to understand the basic Traffic Flow Theory then compare between the theory and measurement result of traffic flow. The traffic flow analysis originally comes from fluid mechanism theory. When average vehicle speed (v) (km/h)on a road and the density of vehicle which is so called Traffic Density (k)(veh/km), the Traffic Volume (q) (veh/h) is obtained by Eq. (1).

$$q = k \times v \tag{1}$$

Figure 2.
Traffic video monitoring camera and high resolution 4 K camera (photo by zero sum ltd).

Figure 3.
Traffic video monitoring camera location.

And the fundamental relationship between (k) and (v) is defined as Greenshields [4] in Eq. (2).

$$v = v_f\left(1 - \frac{k}{k_j}\right) \tag{2}$$

where v_f is free speed which is theoretical maximum speed at zero traffic density and k_j is jam traffic density which is theoretical zero vehicle speed condition. From Eq. (1) and Eq. (2), Eq. (3) is obtained.

$$q = v_f\left(1 - \frac{k}{k_j}\right)k \tag{3}$$

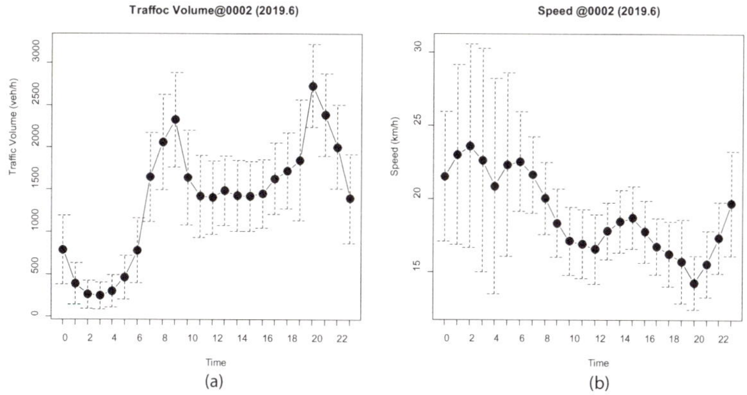

Figure 4.
Example of traffic condition at camera#2. (a) Time-based traffic volume (b) Timeobased vehicle speed.

Eq. (4) is obtained by changing Eq. (3) as traffic density quadratic equation.

$$q = -\frac{v_f}{k_j}\left(k - \frac{k_j}{2}\right)^2 + \frac{v_f k_j}{4} \tag{4}$$

By using Eq. (2) and Eq. (4), we have two funder mental traffic characteristics in **Figure 5**.

From K-Q curve in **Figure 5**, traffic congestion occurs between k_c to k_j because traffic volume Q is decreasing at q_c.

2.3.2 Traffic flow measurement

From measurement data, the fundamental traffic flow characteristics are shown in **Figure 6** as an example of Camera#2 in June 2019.

By comparing **Figures 5** and **6**, the boundary line of each characteristics is similar shape. But the measured data spreads under the boundary line. It is

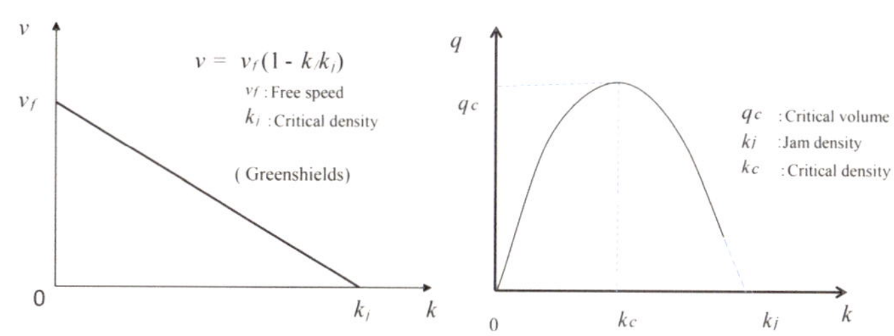

Figure 5.
Fundamental traffic flow characteristics.

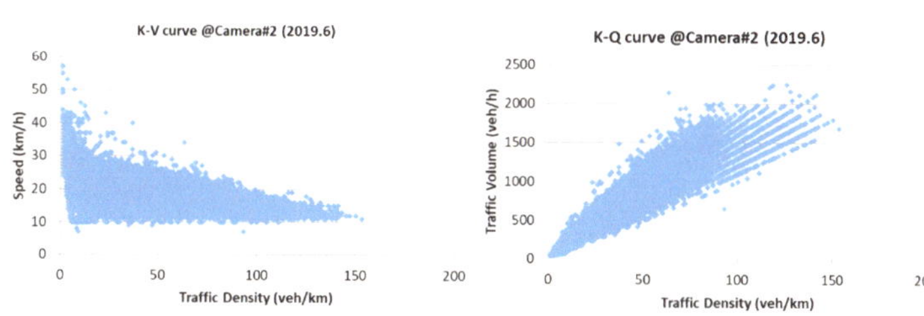

Figure 6.
Fundamental traffic flow characteristics at camera#2.

difficult to figure out the traffic congestion condition from measurement data. We see the traffic congestion occurs around 20:00 at Camera#2 from **Figure 4**. Therefore, the measurement funder mental traffic flow is shown by time zone base in **Figure 7**. There are six time zone from 7:00–10:59 as T1, 11:00–14:59 as T2, 15:00–18:59 at T3, 19:00–22:59 as T4, 23:00–2:59 as T5, and 3:00–6:59 as T6. Then T4 is most critical traffic congestion condition. From **Figure 7**, we see the traffic congestion at Camera#2 in June 2019 occurs the area of the funder mental traffic flow characteristics under its boundary line. This is one of typical features of traffic flow characteristics in India. The traffic congestion occurs before its critical traffic volume (q_c) (refer to **Figure 5** K-Q curve). This research has been done in previous study for time-zone based visualization of traffic flow [5].

When we use the boundary line as its traffic flow characteristics, we get the traffic flow parameter. The example of boundary line for **Figure 6** is shown in **Figure 8**. The we have the following parameter $v_f = 38$, $k_j = 250$ in this case. This is called in previous research as Boundary Observation Method [6].

2.3.3 Traffic congestion

In previous Section 2.3.2, we have traffic flow fundamental characteristics curve and traffic flow parameter i.e. free speed, jam traffic density. In terms of traffic congestion analysis, we only have an observation method by using Time-based Traffic Volume and Vehicle speed shown in **Figure 4**. In this section, we try to get one of some traffic parameters as its traffic congestion condition. In previous research [7], we focus on occupancy (OC) parameter. The occupancy is obtained by Eq. (5) from traffic theory.

$$OC = 100 \times \frac{q}{v} \times \bar{l} \,(\%) \tag{5}$$

where (q) is traffic volume, (v) is vehicle speed, and $\bar{l}$ is average length of vehicle.

Here is an example from our measurement data in **Figure 9**.

Figure 9 shows traffic condition of total traffic monitoring cameras in **Figure 3**. From **Figure 9**, occupancy shows traffic congestion condition well. In case of occupancy as traffic congestion parameter, we are able to see the traffic congestion condition. It is not necessary to use two parameter i.e. traffic volume (q) and average vehicle speed (v) like in **Figure 4**.

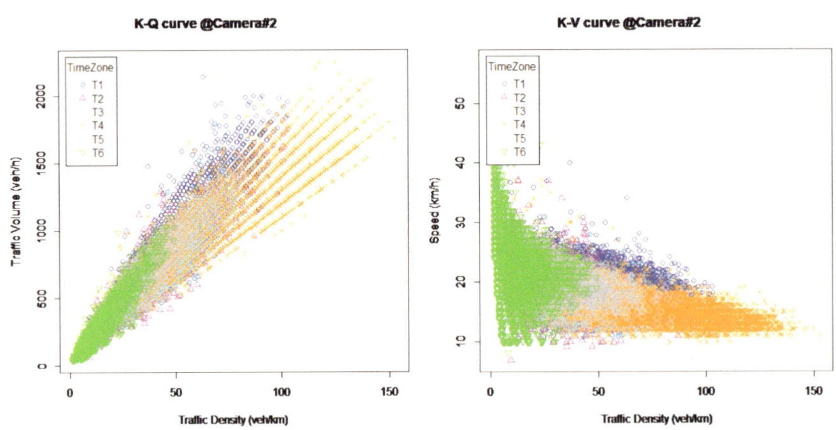

Figure 7.
Time zone based fundamental traffic flow characteristics.

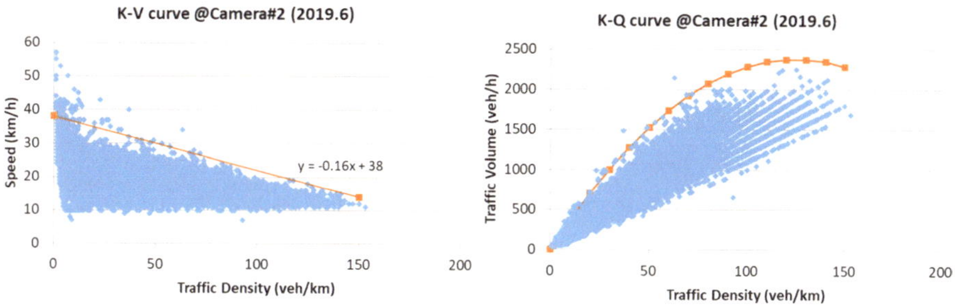

Figure 8.
Traffic flow characteristics with boundary line at camera#2.

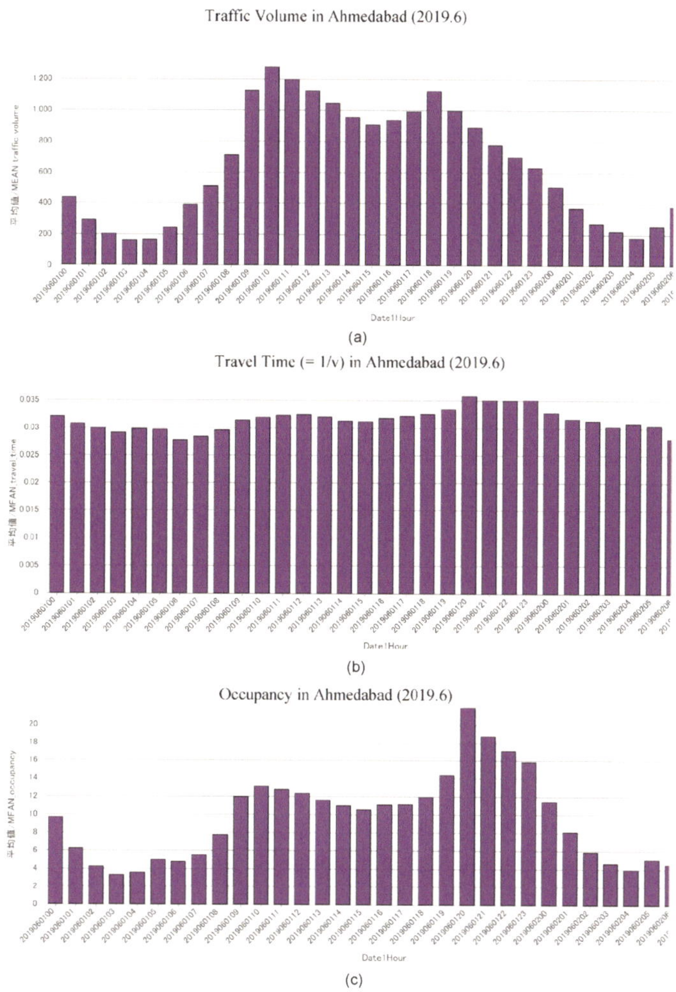

Figure 9.
Traffic flow parameter in Ahmedabad June 1st 2019. (a) Traffic Volume (q) in Ahmedabad June 1st 2019. (b) Travel Time (= 1/v) in Ahmedabad June 1st 2019. (c) Occupancy (OC) in Ahmedabad June 1st 2019.

2.3.4 Traffic congestion condition

The occupancy is useful to understand the traffic congestion condition from the traffic flow parameters in 2.3.2. By using the occupancy parameter in whole city, **Figure 10** shows the trend of traffic congestion condition in Ahmedabad.

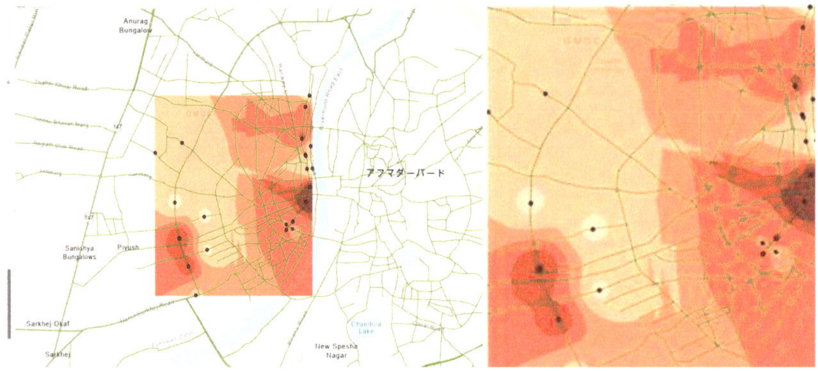

Figure 10.
Traffic congestion by occupancy in Ahmedabad June 1st 2019.

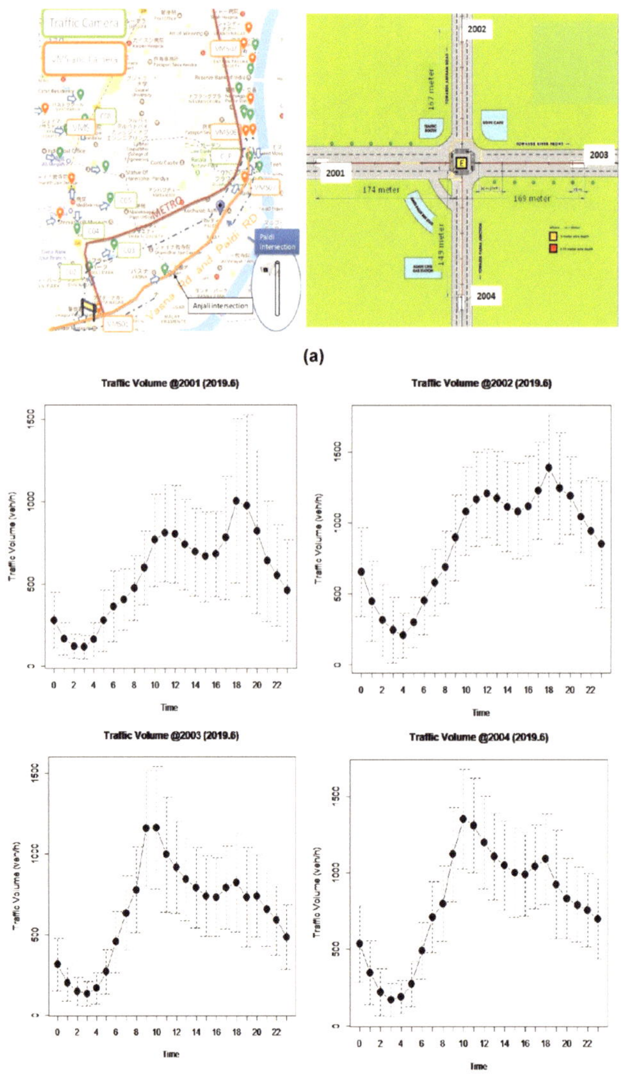

Figure 11.
Padli junction location relation map and traffic volume at Paldi. (a) Location Map (by Google MAP) and Paldi junction Camera position. (b) Traffic Volume at Paldi Junction (Camera#2001 to 2004).

Figure 12.
Drone traffic monitoring at Padli junction.

The data is used all traffic monitoring cameras and interpolation of Inverse Distance Weighted (IDW) with Geographical Information System (GIS) such as ArcGIS.

From **Figure 10**, it is clear about traffic congestion condition in Ahmedabad by using occupancy parameter. The left bottom area if **Figure 10** is crowed area because of shopping center and new business office along the road (132 Feet Ring Road) and the right side of middle area along the road (Ashram Road), which connects with the right side of city which is called "old city", where there are government office Ahmedabad Municipal Corporation (AMC), court, bus garage etc.

In Paldi junction sounded by Camera#2001 through Camera#2004, it is expected to become traffic congestion because the fly-over is under development Ashram Road from Camera#1 to Paldi junction (refer to **Figure 11**). There is METRO under development which will run Ashram Road in parallel (red line in **Figure 11(a)**). In **Figure 11(b)**, there are four graphs about traffic volume at Paldi Junction from Camera#2001 through 2004. From these graphs, there is interesting trend of Paldi Junction traffic. For Camera#2001 and 2002, there is more traffic in the evening rather that in the morning. On the other hand, for Camera#2003 and 2004, its trend is opposite. The direction of each cameras face to the center of Junction, this means many traffic moves from old city or east side of city to new city or west side of city. From these measurements of traffic condition in Ahmedabad, the value of occupancy is lower than 25% in wide area in the city. According to our experience of the project, there is not always congested against our expectation before this project stars. The Ahmedabad traffic congestion occurs by some reason, not always crowed by traffic. This is important evidence and hints how to solve traffic congestion issues in India.

Here is undergoing research activities for monitoring traffic condition. We use Drone at Paldi junction. **Figure 12** shows the video capture at Paldi junction in order to understand vehicle movement behavior. The Drone flied about 10-meter-high at Paldo junction. From the Drone video monitoring, it becomes clear vehicle behavior rather than that by traffic video monitoring camera because traffic video monitoring camera is installed at the fixed point and not high position in the sky. The detail analysis is done in our future work.

3. Traffic safety management

In previous section, we have traffic flow analysis at Paldi junction. From **Figure 12**, we see clear real vehicle behavior by Drone and each four direction of vehicle movement is controlled by traffic signal. In terms of traffic signal in India, the fixed cycle control method is used, which is typical signal control method in all India. There are several traffic signal control such as actuated control type and spatial control type. The actuated control type is to change traffic signal timing based on its waiting length of each direction of road. And spatial control time is to change traffic signal timing synchronized with multiple traffic signals nearby. In case of small number of vehicles, the fixed cycle control type is able to manage. But in accordance with number of vehicle growth, the waiting signal time becomes longer and it makes more que length of each roads. It is necessary to consider their traffic signal control sometime near future.

However there is more fundamental issues in Indian traffic signal. It is relatively quite small installation in major cities. **Figure 13** shows the number of traffic signal installation in major cities in the world [8].

From **Figure 13**, the number of traffic signal installation is quite small compared with other advanced countries, Japan, USA, UK, Singapore etc. In previous research about quantitative traffic safety analysis by Tsuboi.T [9], it is introduced "enhanced Smeed's Law", which is exposition of Smeed's Law. The Smeed's Law is able to explain the relationship between traffic fatality number, population, and number of vehciles [10]. The Smeed's Law equation is provide by Eq. (6) and the enhance Smeed's Law equation is provide by Eq. (7).

$$D = 3\times 10^{-4}\left(np^2\right)^{\frac{1}{3}} \tag{6}$$

where D is the number of fatalities, n is the number of vehicle registrations, and p is the population.

$$D_{t+1} = D_t\left(1+\frac{1}{3}N_t+\frac{2}{3}p_t\right)e^{-\gamma s_t} \tag{7}$$

where D_t, Nt, Pt, and St are the number of fatality, vehicles, population, and signal installation at time t. γ is an exponential constant of installation of signal.

By Smeed's Law and enhanced Smeed's Law, we have comparison results of Indian fatality in **Figure 14**.

Based on Japanese fatality record, **Figure 15** shows each Smeed's Law result and enhanced Smeed's Law result.

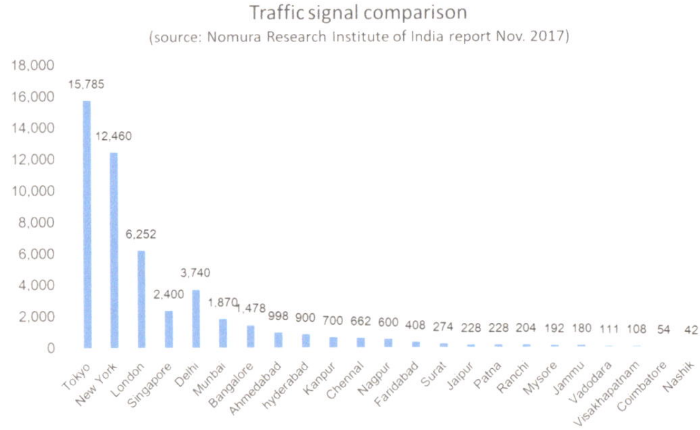

Figure 13.
Traffic signal installation comparison.

When we use γ parameter based on Japanese experience by Japanese traffic safety policy, it is able to expand to Indian fatality analysis in future. **Figure 16** shows the expectation analysis results of Indian fatality under Japanese traffic safety policy.

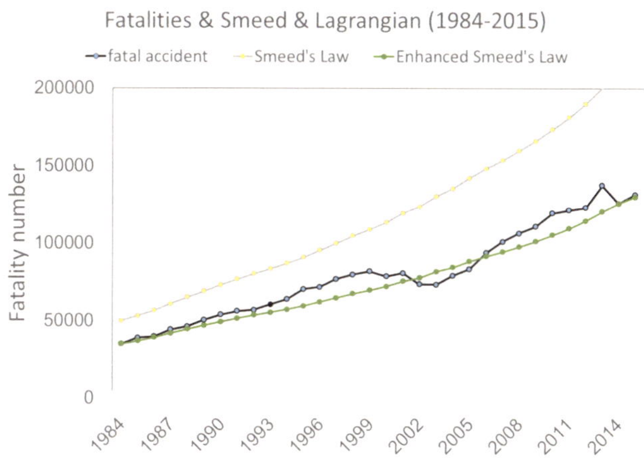

Figure 14.
Indian fatality analysis by Smeed's law and enhanced Smeed's law.

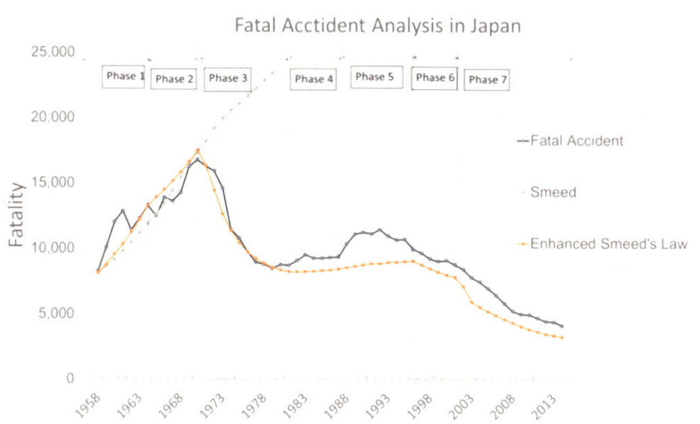

Figure 15.
Japanese fatality analysis by Smeed's law and enhanced Smeed's law.

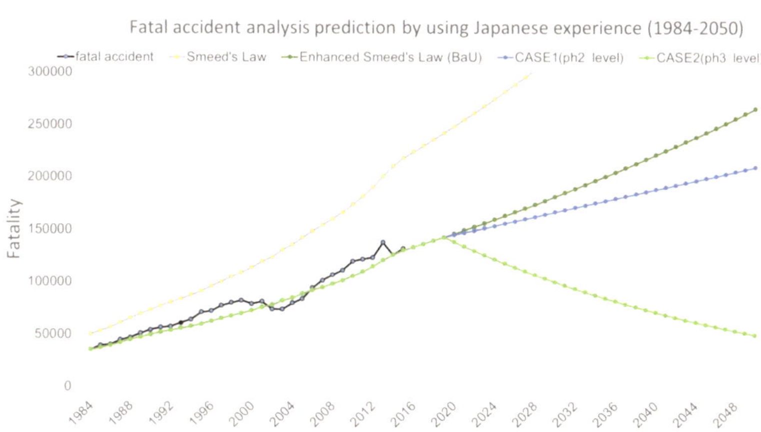

Figure 16.
Indian fatality analysis with Japanese traffic safety policy.

Year	Level	Major action & item	Note
Before 1964	Level 1	Under development for traffic management	
1964	Level 2	1st policy plan	
		Signal implementation	
1970	Level 3	2nd policy plan	
		Vehicle sensing installation	
		Sensitive type traffic signal	
1980	Level 4	3rd policy plan	
		Central control system for signal	Overall Japan
		Control center for signal control	ditto
		Traffic regulation & more installation	ditto
		Education for traffic	ditto
1987	Level 5	4th policy plan	
		Traffic signal algorithm improved	
		Traffic information display	
		Network among centers	
1996	Level 6	5th policy plan	
		ITS system development	
		Congestion control system (VICS)	
		Optimized algorithm (MODERATO)	

Table 2.
Traffic safety policy in Japan.

In **Figure 16**, there are several case studies—Business as Usual (BaU) which means there is nothing special to do with traffic safety, CASE 1 which follows Japanese traffic safety phase 2 level, and CASE 3 which follows Japanese traffic safety phase 3 level. Each level of Japanese safety policy is shown in **Table 2**.

From this case study, we understand that expansion of traffic signal installation is necessary but it is not enough to reduce the number of fatality in India.

4. Summary and discussion

In this chapter, we started how to measure and collect Indian traffic data and visualize its condition with somehow in quantitative value. And we use ordinal traffic video monitoring camera which has been used in worldwide in general. At the same time, we are able to use high technology tool such as drone in this book. But there are other several technologies available day by day like Artificial Intelligent (AI) and Deep Leering so on. But here is important things for traffic flow analysis is that transportation is a kind of human activities. Therefore, we cannot ignore the real people life during this research activities, we have many chance to talk and exchange information with local government and stakeholders of public transportation organization. One of interesting reason of Ahmedabad traffic congestion seems to be occurred in the evening because of people evening activities after their work, it is shopping, dinner with friends, not straight to go home. In current road infrastructure in India, there is no well-developed for transportation infrastructure, one of officer mentioned about public parking space problem. The lack of vehicle parking space in their roads, so when people goes to restaurant in the evening, there is no appropriate parking space, they park their vehicle along the street. Then its road becomes narrow by vehicle parking.

From this story, the traffic congestion is not always by just more vehicles. There is other fundamental reason such as lack of proper parking space in the city. This is unbalance between economic growth and infrastructure preparation. In **Figure 17**, we have four years ago traffic data in Ahmedabad. The condition of each area is not so much changed but just number of vehicles is 1.3 time in **Figure 18**.

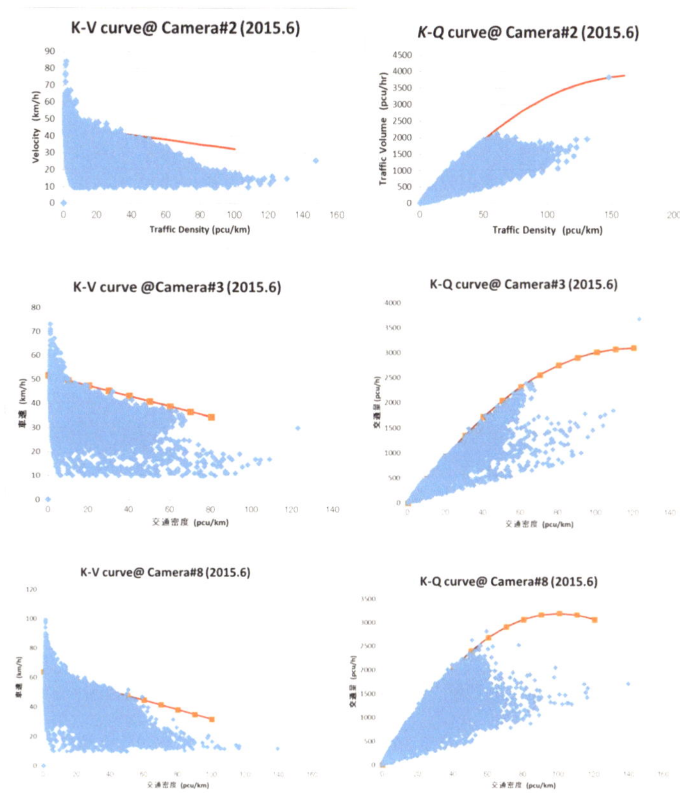

Figure 17.
Traffic flow characteristics in Ahmedabad June 2015.

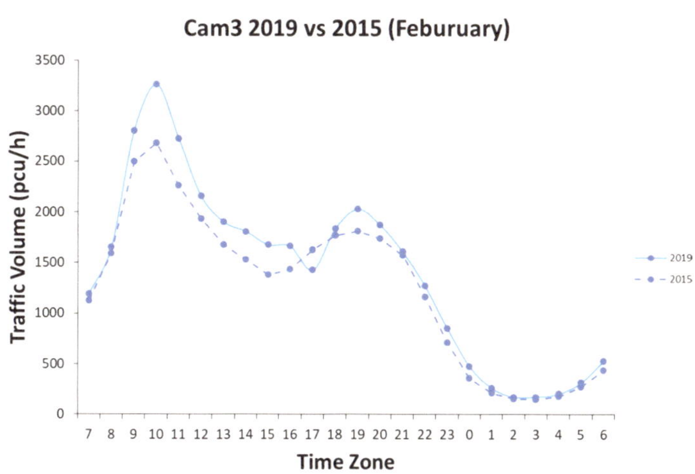

Figure 18.
Comparison time-based traffic volume between 2015 and 2019.

5. Conclusion

In terms of the book title "Megacities – Intelligence, Sustainable and Resilience Built Environment", we see transportation case study in a major mega city Ahmedabad city in Gujarat state of India. And we know traffic issues are critical not only in India but also in any country in the world. The each country difference is based on each countries" condition, economics, social condition, relationship with other cities, states, and countries.

In this chapter, we focus on transportation and pick up one of typical city case study in India. From this case study, the visualization of traffic condition is key especially for understanding real traffic condition in India and it is useful to use advanced engineering technology for sensing traffic condition. We find several reason why the traffic congestion occurs in Ahmedabad city in the case study. From this research, the traffic congestion does not always occur and there is some other reason behind such as unbalance between number of vehicle growth and lack of infrastructure development. This situation is not exception in other advanced countries.

The most important research is to focus on the real human activities, especially transport field. And it is also important to learn other experience of other countries. We see some example for traffic fatality analysis based on Japanese experience. Therefore this kind of research should be shared among the related stakeholders and listen their voice. In some case, it is important that we should understand the target megacities requirement and situation and how to use each experience and share with them. This is not only providing advanced, tools, facilities but our experience in which we have also had similar condition in the past. It enable to us for creating new idea and solution during studies towards our unexpected matters in future.

Acknowledgements

This research is funded by SATREPS, grant number JPMJSA1606.

Author details

Tsutomu Tsuboi
Nagoya Electric Works Co. Ltd., Ama-shi Aichi, Japan

*Address all correspondence to: t_tsuboi@nagoya-denki.co.jp

References

[1] Population of India [Internet] 2020 Available From: https://indiapopulation2019.com/population-of-ahmedabad-2019.html [Accessed: 2020-08-21]

[2] Registered number of vehicles Ahmedabad India FY 2006-2017 Available From: https://www.statista.com/statistics/665754/total-number-of-vehicles-in-ahmedabad-india/ [Accessed: 2020-08-21]

[3] Ahmedabd Municipal Corporation Avilable From: https://ahmedabadcity.gov.in/portal/jsp/Static_pages/introduction_of_amdavad.jsp [Accessed: 2020-08-21]

[4] Greenshields B.D., 1935, A Study of Traffic Capaci, Proc.H.R.B., 14, pp.448-477

[5] Tsuboi T, Time Zone Impact for Traffic Flow Analysis of Ahmedabad city in India, 5th International Conference on Vehicle Technology and Intelligent Transport Systems, May, 2019

[6] Tsuboi T, Yoshikawa N: Traffic flow analysis in Ahmedabad (India), Case Studies on Transport Policy Volume 8, Issue 1, March 2020, pp. 215-228. DOI: 10.1016/j.cstp.2019.06.001

[7] Tsuboi T, Traffic Congestion Visualization by Traffic Parameters in India, International Conference on Innovative Computing nd Communication Jan. 2019.

[8] Nomura Research Institute India report, Nov. 2017.

[9] Tsuboi T, Quantitative Traffic Safety Analysis for India by Japanese Experience, Sociology Study Jan.-Feb. 2020 Vol.10, No.1, pp. 1-14, DOI: 10.17265/2159-5526/2020.01.001

[10] Smeed, R. J. (1949). Some statistical aspects of road safety research. Journal of the Royal Statistical Society, Series A (General), 112(1), 1-34.

Chapter 9

Approach Based on Traditional Architecture Indicators to Strengthen Urban Identity in Diverse Cities

Hatem Ibrahim

Abstract

This chapter presents an analytical study and a review of the contemporary architecture in Doha- The capital city of the State of Qatar. It demonstrates where 'Traditional Criteria" are available in three recently developed cases study. The selected cases study has different values of traditional value that response to contemporary architecture trend. The assessment of traditional value is based on an analytical survey approach. This aim of the analysis is to develop a matrix for unifying local architecture with contemporary architecture in Doha city. The developed matrix will enable local authorities, investors, and architects to assess the value of local architecture language in contemporary architecture for diverse cities in general and for Doha city in particular. The study is valuable to be considered as one of the most attempts investigated work against an ever-increasing loss of local building trend that caused by an industrialized and standardized world. The significant of the study is the demonstration of what is learned and what is gained in understanding contemporary architecture in Doha city. The outcomes include an implementation of a scoring system of different local architecture variable. Future study is suggested to address the design typologies of buildings, and traditional elements to response to the local environment and the socioeconomic of the city.

Keywords: contemporary architecture, diverse cities, traditional criteria, Qatari architecture language, sustainable development

1. Introduction

This chapter contributes a novel approach to strengthen identity in diverse cities. Cities hold unique cultural, social characteristics and represent ideal places for the development of an urban related identity. In addition, a better understanding of urban identity formation and identification processes is essential to comprehend citizens' support or disapproval of current challenging local initiatives [1]. Cities respond differently to migration and their changing place identities. Some cities have been considered as super-diverse cities by defining migration and diversity as one of the drivers that support the economy and the global image [2–4]. Gulf cities show a culturally diverse demography, where international migration contributes to the diversification of urban and this has led to fundamental changes in identity.

Diversity can be considered as urban identity that provides strongly a shared sense of belonging. The diversity is likely to come together with multiculturalist or interculturalist policies towards the verities of culture background [2].

The argument between traditional and global ideas has been quite sharp in some countries and recently the resulting resentment gave rise to fundamentalist movements [5]. In globalization era, diverse cities are encountering great challenges. The term globalization was known during the second half of the 20th century and was not a popular until the second half of the 1980s [6]. The globalization has attracted a significant attention, yet the term is used in many different contexts, many different people and for many different purposes. Globalization as the intensification of worldwide social relations that link distant localities in such a way that local happenings are shaped by events occurring many miles away and vice versa [7]. In that sequence, globalization is considered a discourse of knowledge that elevates awareness of the links between various scales of lifestyle that reflects the diverse of population and cultural background. The Encyclopedia Britannica defines globalization as the process by which the experience of everyday life is becoming standardized around the world [8]. It is a contested discourse showing many variants, some of which are clearly more influential than others and a major force in altering the world's cultural landscapes [9].

The relation between the two trends "local and global architecture" has long existed, with two directions. The first direction seeks to protect and reinforce the traditional building elements in the urban, which aim at the continuity of the history and culture diversity to address identity. The second direction encourages the use of new building elements with the support of technology and developing new materials to meet the functional needs [10]. This chapter endorses Oliver theory [11] through the integration of traditional architecture that support the context of the local environment in Doha city. The theory discusses broad concepts, which is addressing vernacular architecture in diverse cities, such as symbolism, decoration, building materials...etc.

The integration of traditional elements into the contemporary architecture is a popular topic that addressed in many researches. This chapter is continuation of a previous published study by [12]. The study is focused on the theoretical part of the topic by studying the three variables of traditional architecture (Building components, landscape components and function/activities). The findings of this study can be used by the authorities, urban planning decision making and private sectors as a purpose of making the city urban more reflected to the tradition and to embrace modern urban elements in many forms including new construction, renovations and site planning.

2. Methodological approach

The study presents the approach of strengthening the diverse city identity based on native function and activities for Doha city. It answers the following questions:

- What is the definition of contemporary architecture?
- What are the considered factors for addressing traditional in contemporary architecture? and,
- How can the traditional architecture be addressed in different trends of contemporary architecture in diverse cities?

In this study, the following research tools is implemented:

- A content analysis of reference data to establish the different contemporary architecture trends in Doha city. The indicators that support traditional architecture in the contemporary urban also will be identified.

- Empirical analysis by investigating a survey study, which is targeting stakeholders such as experts, residences and tourists.

The objective of the analysis is to evaluate the three established traditional architecture indicators in Doha city: Building elements, landscape components and function/activities.

3. Literature review and background

Alsultani [13] addressed the difficulty in recognizing the contemporary architecture as a clear trend in ether international or regional level. Contemporary architecture was a solid and clear trend during the beginning of the 20th century by emphasizing the function of different urban elements. Antarikananda et al. [14] argued that the internationalism has been reflected in contemporary architecture by scratching the local tradition that has no identity. During this, the traditional elements has ignored despite of its value that basically response to the local environment.

The rapid development of the economic and social drivers throughout the Middle East are more radical now than ever. The region is passing by a transition from the traditional economics and society to the new era in the field of urban planning. This can be noticed during the rapid urban growth that connected with slum area and urban sprawl [15, 16]. Historically, the camps, residences, palaces, and citadels of the ruling class used to set apart from the commoners. This would mean that the larger and more ornate structures used to be as landmark structures and isolated on the edges of major cities away from the urban center. The buildings associated with the working classes, the local crafts workshop and other city services were weaved together in the urban fabric.

The city includes buildings; landscape and the other important related activities that can draw the local identity. Madanipour [17] stated that public elements such as streets, landmarks and open spaces in the city have been considered as a part of the buildings. This is emphasized in Middle East cities, where buildings are connected with the urban fabric and where city neighborhoods are organized their functions and services in traditional trend.

The architecture in Qatar was concerned in protecting building from the hot environment and this trend was negotiated in old architecture [12]. The buildings at this time are to build for last without harming the nature. An example is shown in **Figure 1**, Um Salal Mohammed fort, which is located in Qatar and was built using limestone as a local material.

Since Qatar becomes an open country to the world; it had been affected by the developing different architecture trends. The huge development in Qatar few years ago increased the flow of people with different cultures and life styles. This variety of cultures in addition to the open market and globalization affect the architecture in the country.

Qatar begun to change as a global center for education, media, finance, and culture through several initiatives and large investments such as the Doha Economic Zone, Al Jazeera Channel and Education City. Urban development boom is erected

Figure 1.
Um Salal Mohammed fort in Qatar.

Figure 2.
Contemporary architecture in Doha city.

sport facilities, skyscrapers, iconic museums and gated communities. Now, Qatar is a country where the contemporary architecture can be defined as a mix of European, Asian, Islamic, Arabian and local trends. Accordingly, it is difficult to recognize a clear architectural trend for Qatar in general and for Doha city in particular, **Figure 2**.

4. Three cases study of contemporary architecture in Doha City

The aim of this study is to assess the three indicators that support traditional architecture. This is as an approach for integrating the Qatari architecture language in the urban contemporary environment with the support of city image. Ibrahim [12], identified the three traditional architecture indicators in Doha city "Building components, landscape components and function/activities". This study identifies the value of traditional Qatari architecture in three urban districts in Doha city.

A survey study is developed, taking into account the concluded three indicators that affect the traditional value. The survey study examined the stakeholders' opinion for the previously mentioned three indicators in each district: AL-Dafna, Souq Waqif and Pearl Qatar. The three urban districts have been investigated according to the developed questionnaire. This process is carried out in Ref. to many site visits. **Figure 3** shows the related different questions to each indicator.

The questionnaire targeted different stakeholders: locals, expatriates, tourists in addition to the experts in the field of urban planning and architecture. In the beginning, the three cases study are traditionally ranked by the experts. a) Al- Dafna area is ranked as low traditional value with strong global trend, b) Souk Waqif ranked as strong traditional value and low global trend and c) Pearl Qatar is a mix between traditional and global architecture.

The questionnaire is analyzed for each case study as an approach for implementing the matrix for unifying Qatari architecture language with contemporary architecture. The questions were designed to be close-ended questions and in an easy format for the respondents to answer in short time. It consists of twenty questions that are covered the different three considered traditional indicators. The analysis is based on a comparison of the three cases study.

4.1 The case of AL-Dafna area

Al- Dafna area is a new development district in the waterfront area. It is currently considered as the contemporary financial hub of the State of Qatar. In the late of 1990s many tall buildings have developed in the area, with over fifty proposed buildings to be constructed in future. Al-Dafna was literally planned in 1985 to include a massive land reclamation project along Doha's coastline and started to be developed in the mid-1990s. High-rise office buildings and hotels are shaped the financial district of Al- Dafna along beautifully landscape of Doha Bay in Al- Corniche of Doha. The trend of the urban and landscape components in the area is shown in **Figure 4**.

4.2 The case of AL-Souq waqif

Souq Waqif is a significant touristic area in Doha city. It is renovated for selling traditional products such as spices, garments, souvenirs and handicrafts. Many restaurants also are serving cuisines from all over the world in the area. Despite Souk Waqif is developed for hundred years ago, it has been recently restored back to its original glory. People from different regions are visiting the Souk Waqif to buy traditional goods such as wool, traditional THOBES, jewelry and perfumes. Recently, the Souk is a hotspot for different workshops, hosting several art galleries

Building components: *Evaluate the following building components in terms of Qatari traditional trend.*	**Landscape Components:** *Evaluate the following landscape components in terms of Qatari traditional trend.*	**Function/ Activities:** *Evaluate the following function/activities in terms of Qatari traditional trend.*
* *Rate using the following scale:* *Traditional 1 2 3 Not traditional*	* *Rate using the following scale:* *Traditional 1 2 3 Not traditional*	* *Rate using the following scale:* *Traditional, medium and not traditional*
1. *The Wall* 2. *The Roof.* 3. *The Openings.* 4. The *Structure.* 5. The *Building Materials.* 6. The Ornamentation of Buildings. 7. *The* Form. 8. *The Colour.*	9. Landscape Structure: Landscape composition and configuration in overall. 10. Traffic road Materials. 11. Lighting. 12. Signage. 13. Furnishings. 14. Pedestrian and Pathway.	15. Festivals events. 16. Crafts and souvenirs. 17. Food and drinks. 18. Manufacturing/ industry/workshops. 19. Tours/ transportation/ trips 20. Others please state.

Figure 3.
The questionnaire that been used for the three cases study- AL-Dafna, Souq waqif and the pearl Qatar.

Figure 4.
Urban and landscape trends in Al-Dafna area.

Figure 5.
Building components, landscape components and activities in Souq Waqif.

Figure 6.
Building components, landscape components and activities in pearl Qatar.

and traditional events. The renovation of the Souq Waqif was implemented by using historical photographs shot since the mid-forties and fifties, **Figure 5**.

4.3 The case of pearl Qatar

The Pearl Qatar is a man-made island with area of four million square meters. The Pearl has over 32 kilometers of new coastline. It is developed in a shape of two bays for mix-use "residential, retail and leisure". The island includes 15,000 property units and 45,000 residents. It is developed based on a mix architecture trend of Middle-Eastern, Mediterranean and European culture- offers cosmopolitan charm, outstanding beauty and a contemporary architecture, **Figure 6**.

5. Indicators for unifying Qatari architecture language and contemporary architecture

An analytical study of the three selected contemporary areas is carried out to identify the traditional indicators include: Identifying the historical background of each area, the function involved, the physical elements, the building materials, and physical surroundings. These indicators are implemented along a matrix to support the authorities in defining the value of traditional architecture in the contemporary urban.

5.1 Building components

Responses to the questionnaire show that no significant building components in Al-Dafna area that reflect the traditional. Most of the answers conclude that the wall, roof, opening, structure, form and building color are not reflecting the Qatari architecture. However, more than 50% answers conclude that building materials and ornamental elements in Al-Dafna area has a medium Qatari architecture value.

In Pearl Qatar, the percentage of the responses are competitive regarding the structure, building materials and the building form, but the higher percentage concluded that these elements are reflecting the Qatari architecture in a low value. Most of the responses also concluded that the wall, roof and the opening are reflecting the traditional architecture in low value. However, the ornamental elements and the colors of the building are reflecting the local architecture in a medium value.

In the case of Souk Waqif, all variables of building components were given by the highest score in terms of being highly traditional value.

The comparison between the three cases study in terms of the building components is shown in **Figure 7**. The analysis concludes that Souq Waqif is the most significant area that reflects the Qatari architecture for all building components, followed by Pearl Qatar for most of related variables.

5.2 Landscape components

In Al- Dafna area, the landscape structure and lighting variables have a low traditional value. However, the traffic road materials, signage, furnishings, pedestrian and pathway reflect the Qatari architecture in a medium value. There are no landscape components in Al- Dafna area that is reflecting the local architecture of Qatar.

In Souq Waqif, the furnishings, signage and pedestrians and pathway are considered as high traditional value that emphasizing the Qatari architecture. However, traffic road materials, landscape structure, and lighting are considered as average value.

In Pearl Qatar, the higher percentage of the responses concluded that the landscape components is a medium traditional value.

The comparison of the Landscape components of the three areas is shown in **Figure 8**.

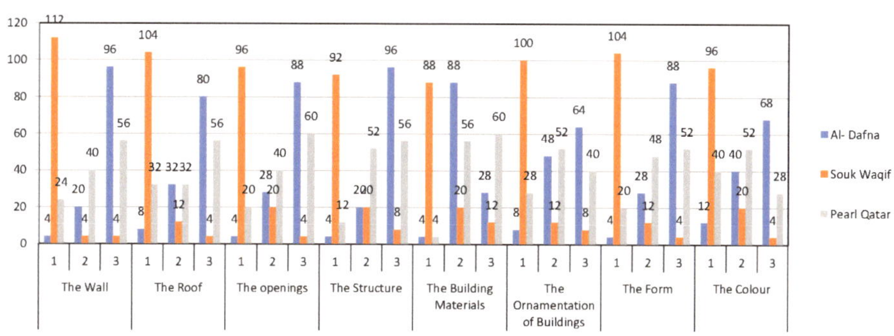

Figure 7.
A comparison between the three areas (Al-Dafna, souk wakif and pearl Qatar) - building components.

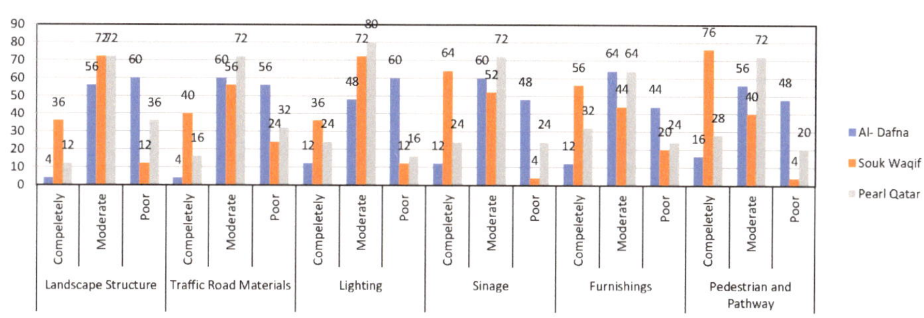

Figure 8.
A comparison between the three areas (Al- Dafna, souk wakif and pearl Qatar) - landscape components.

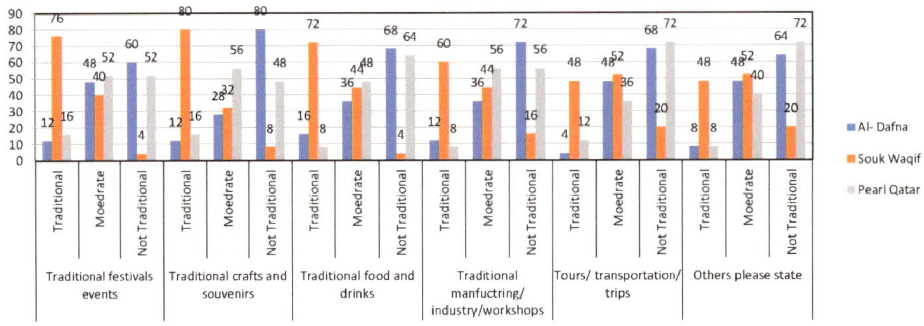

Figure 9.
A comparison between the three areas (Al- Dafna, souk wakif and pearl Qatar) - function/activities.

5.3 Function/activities

The function and activities in Al- Dafna are concluded as a low traditional value. The area has activities that reflect traditional elements such as: traditional tours and events in addition to some transportation trips.

Souq Waqif has many function and activities that emphasizing the traditional trend. The tours and transportation trips have concluded as a medium value.

The function and activities in the Pearl are not reflecting traditional value and this is reduced the traditional value in the area.

The comparison of the function and activities in the three cases study indicates that Souq waqif is the highest traditional value area for the different variables of the function and activities as shown in **Figure 9**.

6. Verification and developing traditional responsive contemporary architecture

In reference to the previous analysis, the findings can be presented in a matrix as shown in **Figure 10**. The different traditional variables for the three indicators are categorized into three values in emphasizing the traditional architecture - High, medium and low.

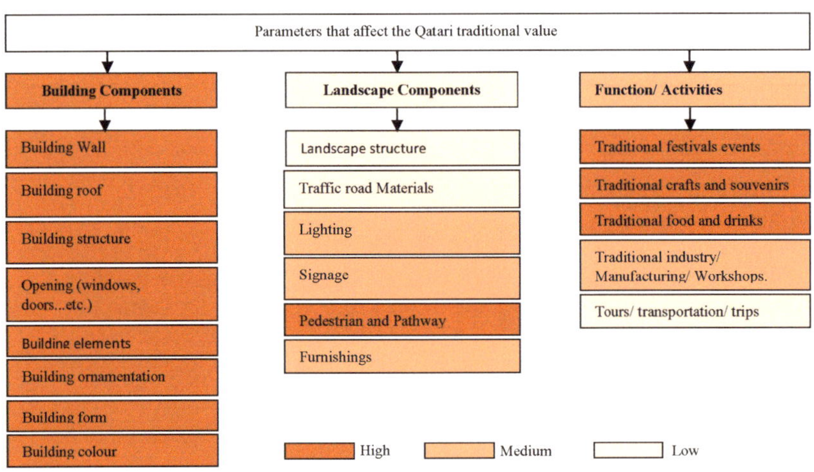

Figure 10.
The ranks of the traditional indicators in emphasizing the Qatari architecture language.

The building components are the highest ranked indicator that support the traditional trend in the area. This is including the wall, roof, structure, opening, building material, building ornamentation, building form and the building color. The integration of the previous mentioned variables can strongly emphasize the traditional architecture trend in the area.

The pedestrian and pathway as one of the variables of landscape components in Souq Waqif is emphasizing the traditional value. However, the landscape structure is traditionally having a low value for addressing the traditional architecture. The signage and furnishing are having a medium value in emphasizing the traditional architecture. The lighting as one of the landscape components has a medium value in the areas of Souk Waqif and Pearl Qatar.

Traditional festivals events, crafts and souvenirs in addition to traditional food and drinks are concluded as high traditional value variables as a part of the function/activities. Tours and transportation trips have a medium value in emphasizing the traditional architecture trend in Souk Waqif and Al-Dafna areas.

7. Conclusion

Urban planners and stakeholders are facing the phenomena of addressing the city identity. This is important to be considered to confront new globalization circumstances. It is required to combine the building technologies while addressing Arab architectural heritage, to absorb and assimilate foreign cultural influences without losing the city identity. This can be addressed by encouraging designers and policy makers to emphasize the dominant traditional value in the region in order to achieve a degree of consensus among Arab intellects.

This chapter focuses on the case of Doha city. It defines the importance of dealing with the diversity and traditional urbanism in contemporary architecture. It aims to establish a matrix for integrating the traditional architecture into contemporary architecture. In this context, the study firstly provides a conceptual understanding of Qatar traditional architecture; secondly, it provides a matrix of weighting contemporary approaches based on three cases study of contemporary urban areas in Doha city. The considered indicators of Qatar traditional language include building components, landscape components and function/activities. A survey study is developed by Ibrahim [12] and defined the different traditional variables under each indicator. The survey study is considered Qataris, expatriates in addition to the expert in the field of urban planning and architecture. An analytical study based on the findings of the survey has been conducted for each case study.

The building components is basically the elements of building envelope, include roof, walls, windows, skylights, etc. According to the study, the building components are the highest indicator that reflect the traditional value in contemporary architecture. Accordingly, the building envelope must be properly designed by using the traditional elements. It is argued that the traditional landscape components that reflect the culture of the country, can offer a good traditional value in the contemporary architecture. Traditional landscape variables such as: structure, road materials, signage, furniture, etc. has a good role in addressing the traditional value in the contemporary architecture. The traditional function and activities are emphasizing the urban space and has an important role in addressing the traditional value in the contemporary urbanism indirectly.

The findings of this study are based on verbal interviews, questionnaire and observations study that carried out in the selected cases study. The study developed a manifested potential for the integration of the traditional value into the contemporary architecture in Doha city. This is to provide a balance between traditional and diversity in multi-cultural cities, reinforcing a distinct national identity while at the same time remaining open to diversity influences.

The study contributes in empowering the identity in diverse cities in general and in Doha city in particular. It is bridging the relevance gap between theory and practice in the urban planning. The findings also enable to support a future developed sustainable urban by the integration of the developed identity indicators into the planning processes.

Acknowledgements

This study was made possible by NPRP grant number [NPRP 07 - 960 - 5 - 135] from the Qatar National Research Fund (a member of Qatar Foundation). The findings achieved herein are solely the responsibility of the authors.

Author details

Hatem Ibrahim
Department of Architecture and Urban Planning, Qatar University, Doha, Qatar

*Address all correspondence to: hatem_ibrahim@qu.edu.qa

References

[1] Belanche, D., Casalo, L. and, Carlos Flavia, C (2017). Understanding the cognitive, affective and evaluative components of social urban identity: Determinants, measurement, and practical consequences, Journal of Environmental Psychology 50 (2017) 138e153

[2] Belabas W, Eshuis J, Scholten P. Re-imagining the city: Branding migration-related diversity. European Planning Studies. 2020;**28**(7):1315-1332. DOI: 10.1080/09654313.2019.1701290

[3] Glick Schiller N, Çağlar A. Towards a comparative theory of locality in migration studies: Migrant incorporation and city scale. Journal of Ethnic and Migration Studies. 2009;**35**(2):177-202. DOI: 10.1080/13691830802586179

[4] Hassen I, Giovanardi M. The difference of 'being diverse': City branding and multiculturalism in the 'Leicester model. Cities. 2018;**80**:45-52. DOI: 10.1016/j.cities.2017.06.01

[5] Saqaaf A. The Middle East City: Ancient Traditions Confront a Modern World. New York: Paragon House Publishers; 1986. p. 6

[6] Holm I. *Ideas and Beliefs in Architecture and Industrial Design*. PhD thesis. In: AHO – The Oslo School of Architecture and Design. 2006

[7] Oncu A, Weyland P. Space, Culture, and Power: New Identities in Globalizing Cities. London and New Jersey: Zed Books; 1997. pp. 1-2

[8] Adam R. Globalisation and architecture: The challenges of globalization are relentlessly shaping architecture's relationship with society and culture. The Architectural Review. 2008;**223**(1332):74-77

[9] Kobayashi K, Jackson S. Globalization, creative alliance and self-orientalism: Negotiating Japanese identity within Asics global advertising production. International Journal of Cultural Studies. 2019;**22**(1):157-174

[10] Lewis R (2002). Will forces of globalization overwhelm traditional local architecture? Washington Post 2, 20th November 2002

[11] Oliver P. Encyclopedia of Architecture of the World. Cambridge; 1997 9780521564229

[12] Ibrahim H. New paradigm of urban development: Envisioning sustainable futures in Qatar, proceedings of the 6th international conference on sustainable development and planning, Wessex institute of technology. In: 27-29 May. Vol. 2013. Greece: Kos Island; 2013. pp. 299-310

[13] Alsultani, K. (2011). What is contemporary architecture? Available: http://omranet.com/vb/showthread.php?t=3506&page=1

[14] Antarikananda P, Douvlou E, McCartney K. Lessons from traditional architecture: Design for a climatic responsive contemporary house in Thailand, proceedings of the. In: 23rd Conference on Passive and Low Energy Architecture- PLEA2006 -, Geneva, Switzerland, 6-8 September 2006, Pp. 2006. pp. 11-43

[15] Bianca S. Urban Form in the Arab World: Past and Present. New York: Thames & Hudson; 2000

[16] Lapidus IM. Middle Eastern Cities. Berkeley: University of California Press; 1969

[17] Madanipour A. Design of Urban Space: An Inquiry into a Socio-Spatial Process. Tehran: Processing companies and urban planning; 2008

Chapter 10

Architectural Design Canons from Middle Ages and Before: An Inspiration for Modern Sustainable Construction

André Frans De Naeyer

Abstract

The role of geometry and arithmetic in ancient building is common knowledge, but it has seldom been proved by measured drawing. This chapter looks for the remote origins of design criteria and ancient canons, and their application in representative antique and medieval architecture. Architectural design had to reflect the universal cosmic Order and Harmony and the ancient and medieval architect-designer had to rely on the same intangible instruments, i.e. the geometry and the arithmetic's, created by the Divine Geometer. The geometry of forms and the numbers of quantities and dimensions served as a mayor instrument for developing coherent modulation in the design and the structure of the building and his environment. They also served as a symbol and an allegorical sign to convey intangible messages from the commissioner. Metric analysis reveals this evident design practices and their probable semantic content. This is illustrated in the analysis of six cases: the Cheops pyramid at Memphis, the Pantheon at Rome, the Charlemagne's Palace Chapel at Aachen, the Our Lady's Cathedral at Chartres, the S. Francis Basilica at Assisi and the Castel del Monte at Andria. This historic examples should inspire modern creative design and modern sustainable construction.

Keywords: metric analysis, modulation, medieval design, design arithmetic and geometry, design history, ad quadratum, proportion, golden mean, sustainable construction, Cheops pyramid, Pantheon Rome, Carolingian Chapel Aachen, cathedral at Chartres, S. Francis Basilica, Castel del Monte Andria

1. Introduction

The study of architectural history is and always has been one of the fundamentals in each architect's education; up to the 1920's circa also the design classes were embedded by the replica of the Greek temple orders and the imitations taken from ancient reference books such as Vitruvius, Palladio, Viollet-le-duc and many other. This changed after the 1920's in favor of a more multidisciplinary and technical orientated training, promoted by "Ornament and Crime" (A. Loos, 1908), "Form follows function" (R. Sullivan, 1900 ca.) and other design maxims. It caused a proportionate impoverishment of building history knowledge and of traditional design methods. One said the architect-artist's mind had to be liberated from all historic ballast, and should be able to create within his free individual creativity. This is a discussable principle with potentially quite negative consequences as architecture is not only a question of artistic creativity or aesthetical harmony, nor a pure functional or technical discipline: "Architectura ... nascitur et fabrica et ratiocinatione ..." ("Architecture is born by craftsmanship and balanced rationality", Vitruvius, I°sec. b.C.) in ([1], I,1). Architecture (with capital) needs both approaches and apart from the Vitruvian "utilitas, venustas et firmitas [1]", Architecture always had an existential and universal dimension dealing with bringing sense and structure in the surrounding space, including physical communication with meanings and messages' to the observer [2].

The often poor knowledge on historic design criteria nowadays, inevitably leads in many cases to a considerable loss of 'sense' and a different type of 'meaning and message' in contemporary projects. Many heritage buildings get their conservation status because of tangible cultural and historic characteristics, and in many cases it is completed with a large intangible content expressed through symbolisms and allegories. Unfortunately, very often this symbolism and allegories get lost today as man is not familiar any more with the ancient allegorical languages. Also the other way around, modern design rarely uses those so called 'old-fashioned' allegorical indications in such speechless but most effective communication between designer and observer. Medieval buildings are particularly representative for the presence of this mostly forgotten intangible communication content, expressed through the symbolism of form, number, proportion, material or color. Based on the analysis of some representative medieval buildings, this chapter illustrates and tries to detect such design indicators to inspire the contemporary designer, not suggesting a flat imitation but a personal modern interpretation and use of the very same ancient design indicators. The two mayor instruments to all kind of allegorical allusions in medieval design are the geometry of the architectural form and the arithmetic's within the different quantities and dimensions.

This book aims contributing in sustainable construction. The easy re-use or reconversion without great structural change or loss of architectural identity is part of all sustainability and certainly one of the most crucial assignments today. Recent experiences on the reconversion of existing fabric or the recuperation of ancient abandoned structures, mostly for evident economic reasons, have proved abundantly that reconversion or recuperation is much easier and less invasive with ancient well-modulated traditional buildings as it is the case with some contemporary building or probably shall be with one of the super eye-catching designed ones, created by great archistars as e.g. the Bilbao Guggenheim Museum or the Baku Heydar Aliyer Center. Certainly, those superlatives are strong signs of digital design and technical knowledge, but their quality remains one-sided, limited to never seen forms and materials. They do not show great flexibility nor long lasting esthetic pleasure; any probable later intervention, as proof of sustainable (re-)use, risks to damage considerably their actual identity. Society needs avant-garde, but this has to be applied with cure and caution. Contemporary design should reconsider the historic canons, take profit of the three thousand years' experience, evaluate and integrate the old principles for harmony and sustainable use in the modern design algorithms to guarantee qualitative architecture and long lasting construction.

2. Ancestral origins of numbers and geometries in West-European architectural design

2.1 The ancient Greek law on measured and figured numbers (Pythagoras and the Pythagoreans)

Unlike what many people thinks, architectural design is not so much a question of spontaneous creativity but much more of theoretical and technical knowledge.

"Ars sine scientia nihil est" according the well know exclamation of the French master builder Jean Mignot consulting the Milan Cathedral builders in 1400 ca. The theoretical "scientia" at Mignot's time was little and approximate, the artistic drive on the contrary was all the stronger. The success of so many ancient buildings, in particular the audacious finely jointed gothic structures were the result of practical experience during about 3000 years of building since the first Mesopotamian temples. Those ancient buildings were always expression not only of specific needs but also of the then living spiritual concepts about society, religion and esthetics. As in many other disciplines, also architecture and design owe a lot to ancient Greek philosophers from the early 10th century BC (and the Egyptians before them) as founders of West-European culture. Within the larger context of the Mediterranean Basin they developed a world view, not precisely as told in Genesis, but quite similar, i.e. created by a Supreme Divinity who organized and structured the initial chaos using calculated and measured geometric forms. This cosmos of a well ordered celestial and terrestrial creation by the Divine Geometer was the example that man had to follow in structuring his own small local chaos of space. All architectural project implies structure of space, and for that reason, all architectural design must be based on calculation, arithmetic and geometry. This idea was further developed, especially by the Christian scholastics (ca. 9th–13th century) and became an existential obligation for all architectural projects. This explains the permanent presence of numbers and geometries in architectural design for more than 3000 years. In this prospective, one could consider Plato, Aristotle, Pythagoras and Euclid (and the unnamed Egyptian and Mesopotamian priests) as the founders of European design principles.

To study the cosmos's structure, Greek philosophers developed arithmetic number systems and geometric procedures to explain the phenomena of life and nature [3, p. 7]. Numbers are abstract concepts related with the quantity of things, but in relation with real sets or groups they become a tangible reality which, in ancient metaphysic thinking, got very often an intangible connotation or symbolic value, variable in history. In architecture, this number symbolisms got related with the physical quantity of distinct built elements or with the measured quantity of length, width, height or volume. Also the procedure to establish a single number, i.e. the type of calculus by simple arithmetic using the four basic operations (addition, detraction, multiplication and dividing) or more complicate ones (square or cubic root), and the position of each single number as part of a sequence (arithmetic, harmonic or geometric progressions) got specific symbolic meaning and became associated with human or natural phenomena or events. In particular the 'harmonic progression', i.e. where each number of a simple sequence stays in 'harmonious' proportion to the previous and the following number, were popular and looked for[1]. The numbers, visible in real quantities (e.g. number of piers, of bays, of rooms, of corners, of stair-steps, ecc.) were a fundamental part of each building design; the same numbers served as the metaphoric indicator par excellence for expressing intangible values or messages such as power, devotion, glory, utility, science, beauty, harmony and other.

This is not the place to enter in detail about the number systems and the numeral calculus, nor about the wide range of symbolic values in Greek and/or medieval

[1] The neo-platonians distinguished ten types of 'harmonic' sequences (including the progression in XIIIth century called after Fibonacci) as those of at least three consecutive numbers of which the proportion between the central number and the previous and the following number 'sounded' particularly harmonious and rhythmical, conform with the essentials of the Pythagorean harmonic canon – the numbers indicating the respective length of each string in playing the Greek lyre, e.g. 1–2-3; 2–4-5; 3–5-8; ecc. [4], p. 43] [5], I,16.

numbering. This chapter only stresses their presence and application since ancestral times, and their fundamental role in the genesis of all pre-industrial building projects. Understanding the ancient metaphors, hidden behind the physical quantities and dimensions in the building, is not so easy as the correct lecture and interpretation of the dimensions presumes the often missing knowledge about the metric unit (yard, foot, cubit?), about the eventual modulus (fixed group or set of units) and about the measuring and building conventions at the time and the place of the design. On top of this uncertainty, the modern observer is seldom familiar enough with the ancient design canons and number or figure symbolisms. The Pythagoreans (IVth-IIIth century BC) knew many types of numbers: real or rational ones, integers, fractions, even and uneven ones, primes, perfect numbers[2], as well as irrational and complex ones (roots, unlimited ratio's such as π (= circonference : diameter of cercle = 3,14 ...) or φ (golden mean = 1,618 ...), and numbers with virtual connotations (sacral, male, female[3]) and still other types.

Number 'one' is seen as the most important number, being the origin of everything, not only in arithmetic calculation but also in the natural world and the cosmos (also the justification for monotheism; although many cults worshipped a Divine Threesome in one Union, i.e. the Holy Trinity in Christianity). Number 'two', first and only even prime, represents dualism, the base of philosophy and all science; number 'three' means the female and number 'four' the male element in the 3–4-5 triangle. Number 'four' also refers to all groups of four elements in nature: the basic elements of everything (earth, water, fire, air); four cardinal directions, four seasons and, in Christian context, e.g. the four evangelists). The sum of these first four initial numbers 1 + 2 + 3 + 4 gives the number 10 (the sequence called "tetractys"), creating the sacral number 'ten', representing the universal order. Because of this special property, 'ten' got a special 'mystic' value and the Pythagoreans cultivated a particular preference for decades and pentades in arithmetic calculus and their homonymous polygons in geometry. The tetractys sequence generated the concept of calculated harmony in a eight-divided music-scale (from second to octave, the double of four tone intervals[4]), and the ancient eight-divided foot unit as well as the modern decimal measuring system. Also Vitruvius, explaining and defending the use of anthropomorphic dimensions, presented the number 'ten' as a sacral and most 'beautiful' number ([1], III,275)[5] . The theory and philosophy on the use and allegorical value of numbers in ancient times is large and filled with unexpected results, but their decisive role in pre-industrial design and sometime also in post-industrial projects, is evident.

The most curious invention from ancient Greece, without any doubt, regards the concept of 'figured numbers'. This means that the number (except number 'one')

---

2 A 'perfect' number is each positive integer which is the sum of all his divisors except itself (e.g. 6, 28, 496, 8128; there are only this four perfect numbers under one million).

3 The gender-ification of some numbers goes back to Plutarch (1st century a.C) who tells in his book *Isis and Osiris* that the vertical side of each Pythagorean orthogonal triangle (i.e. of which the length of all three sides sign a integer number, e.g. 3–4-5; 5–12-13; or 7–24-25, ...) is considered a male element and the horizontal base is the female element; the hypotenuse was seen as the product (the child) of the union of both cathedes.

4 The tetractys property has generated the 'Pythagorean Canon' in music theory, comparable with the geometric proportion (i.e. the length of the Greek lyre strings) and the acoustic harmony in sound proportion or symphonic composition [4], pl.XXXV

5 Also Vitruvius connected the use of numbers with the origins of the cosmos created by God. 'Ten' was considered the most 'perfect and sacral' number as man, created after God's image, had ten fingers to serve the Lord [1], 1,II,27.

should not be seen as a single independed entity, but as a set or distribution, or as a part within a progression, and can be represented in space (linear, superficial or volumetric). The abstract number indicates the ratio between a certain quantity and the unit or dimension of that quantity on which it is relying (in this case on two-dimensional figures or surfaces or three-dimensional volumes). The philosophical background of the concept is more complex and relies on Plato's theory on the proportions of volumes in the dialog Thééthète and presumes the alchemic mixture of arithmetic and geometry [4], p. 45. The concept of 'figured numbers' is particularly useful and explanatory in case of irrational numbers such as root $\sqrt{2}$, $\sqrt{3}$, $\sqrt{5}$ or the real quantities $\pi = 3{,}14 \ldots$ or $\varphi = 1{,}618 \ldots$ as this are infinite ratios. For example, when it is impossible to write the result of $\sqrt{2} = 1{,}41421\ldots$ as a complete and absolutely correct cipher as the result is infinite, the same quantity can perfectly and correctly be indicated and represented in space as the finite length of the diagonal of a square with the side equal unity. In such context, $\sqrt{2}$ is called a 'figured number' as it is associated with the finite length of this line. This is quite important concerning the 'measurability' of the building and admits the integration of root-proportions (most popular in medieval design where $\sqrt{2}$, $\sqrt{3}$ and $\sqrt{5}$ appear frequently) in the design without creating the feeling of approximation or ambiguity (although the tracing of infinite ratios did not create any practical problem at the building site as all dimensions were traced using compass and not with measuring rod). All numbers indicate an abstract quantity which had to be measurable and made tangible in space by length, height or volume. This double character (arithmetic and geometric) of the 'figured' number or ratio is the reason and the instrument at the same time for the presence of the polygons in plan and elevation of medieval buildings; together with the number symbolisms they expressed different kind of allegories or hidden messages, being understood only by the initiated members[6]. For the modern observer, it is always an intriguing challenge to discover the hidden symbolisms in ancient buildings.

The combination of architecture and number philosophy has nothing to do with "numerology", being a predominant esoteric discipline of fortune-telling and kabalistic or astrologic reading of phenomena about man and nature. It does not apply the scientific and rational 'theory of numbers' as intended by Greek philosophers, although even they did not use always the most objective logic, as e.g. by naming male and female numbers, inherited from Egypt. Part of this "numerology" is the practice of the old Hebraic 'gematric'-modus (i.e. giving a numerical value to each letter of the alphabet, making it possible to convert letter-words into a mathematical value), used sometimes in the design of mayor buildings but forbidden in church design by the ecclesiastic authorities.

Finally, the rather primitive measuring instruments and the long lasting construction programs, forced ancient building practice to use preferably integer quotes and simple fractions (half, quarter, third), to facilitate tracing and execution on the building site. This explains the preference for integer numbers in the design of plan and elevation of a building. One also has to consider metric rounding after theoretic calculation and the difference between theory and practice to facilitate execution. Such condition on top of the normal building tolerances, on top of the physical degradation and deformation of historic buildings, ask for benevolent interpretation margins.

[6] One should remember that architectural projects were created by highly educated officials at the service of King or Bishop, and in this quality they were very well aware of the current philosophical and technical achievements of that time.

2.2 Euclid's geometry, the basic instrument in structuring and meditating on space

The geometry is indispensable in structuring any chaotic space. Similar with the procedures used by the Divine Geometer, also man had to create order and harmony by using appropriate geometric figures and proportions. Euclid of Alexandria (ca. 325–265 b.C.) wrote the first systematic manual on this matter, and from that period, a large gamma of regular and irregular geometric forms was developed. The numeric quantity and form of the angles and sides of the figures, in combination with other geometric properties as size, symmetry, congruency, similarity or opposition, they got special symbolic meaning in their architectural application. The most evident figures used in architectural design are the different types of lines (strait, bowed, dotted, alternated), the regular bi-dimensional figures (square, circle, triangle, polygons), and their tri-dimensional derivate. Plato's description in Timaeus on the symbolic content of the regular polyhedra found many applications: the tetrahedron (fire), cube (earth), octahedron (air), dodecahedron (heaven with 12 constellations), and icosahedron (water) [3].

The wohltemperierte amalgamation of geometry and numbering were the necessary conditions for all harmonious architecture; they were the real determinants and driving forces in the design process, and the real generators of all architectural styles. Everything must be calculated, measured and proportionate, as the Holy Bible's verse "Omnia in mensura et numero et pondere disposuisti" (Thou hast ordered all things in measure, number and weight – Book of Wisdom 11:21). This also explains the frequent presence of specific proportions such as the 'Golden Mean' or 'Divina Proportione φ' (= 1,618) not because it was seen as a particular beautiful proportion, but because it was a unique and exclusive value obtained through division of another number (or length of a line) by the 'extreme and mean ratio' or 'mean proportional' and could be seen as a squared figure or 'figured' quantity . The success of this 'golden mean' was also connected with the quite simple procedure to draw it with the compass. It got a particular semantic content as being the irrational, infinite quantity, related with the 'figured' pentagonal number 'five' in the proportion between the diagonal and the side of the pentagon ($\varphi = \frac{1}{2}\left(1+\sqrt{5}\right)$.

As said, the most frequent geometries were the circle, the square and the triangle, as this were the most easy figures to draw up with simple instruments as wooden rod and cord, compass and plumb, but also because of their specific semantics generated since ancestral times. Before Columbus (ca. 1492), the image people had about the structure of the cosmos was that of a flat and square earth (with Jerusalem in the center) and a celestial half globe. It seems logic that the square and the circle, representing the earth and the heaven, were the first geometric figures used in architectural creations. Plato's vision on the origin of the cosmos and the 'elementary triangle' as the fundament of all matter, together with his exaltation of mathematics and geometry at the expense of artistic creativity, contributed considerably in the use of different kind of triangles and the five polyhedra. Christian philosophers extended the ancient semantics with biblical or religious connotations as e.g. the circle became the representation of human society with God in the center and the people staying on the circle line, equidistant from the center and meaning that everyone is equally considered and protected by God. The square represented the walled Terrestrial Paradise or the walled terrestrial and celestial city of Jerusalem.

Apart from the circle, the triangle (equilateral, isosceles, rectangular, proportioned) and the four-angle polygon (square, rectangle, parallelogram, trapezium and rhomb) are the most frequent figures in architectural design, because of their large semantic spectrum and the easy designing technique. For that reason, they are the most popular geometries in architecture. The design 'ad quadratum' i.e. using different squarely formats connected, turned around, divided or superposed, was very popular in all kind of utility-building, 'ad circulum' was frequently used in centralized buildings (e.g. sepulchral monuments, baptisteries); 'ad triangulum' was most appropriate for the design of the building elevation and applied in many church buildings. Also in the panoptic of triangle-types, all had his specific symbolic meaning related with their arithmetic and geometric properties, e.g. the equilateral or 'perfect' triangle (symbol for the divine Trinity: three gods equipotential, united in one figure), the rectangular Pythagorean triangle with figured numbers on each side and, with female base and male height; the isosceles triangle symbolizes Christ: divine and human at the same time. Also triangles with specific ratios were used, as e.g. the 'Egyptian' triangle (as it signs the profile of the Cheops pyramid) is a isosceles triangle with the height equal to 0,625 (= 5/8) of the base (the most beautiful triangle according to Plutarch). Similar semantic discours got connected with all mentioned polygons (pentagon, hexagon, octagon and other).

3. Vitruvius and 1.500 years of modulation through numbers and geometries

As numbers and geometries are the mayor determinants in free architectural design (i.e. without conditions from the commissioner, from materials, or from topography), the definition of the right number and the proper geometry will determine the quality and the legibility of the final product. The correct selection and the proper combination of both determinants within the context of a given assignment, signs the art and the discipline of good design. Design is a research activity, similar at all other scientific research, and this chapter has not to enter in research methodologies or procedures, but intends to look after those tangible criteria used by earlier generations. Moreover, every designer, working on a specific commission, sitting in front of an empty piece of paper (or a white computer display) knows very well the process of trial and error, characteristic for all design processes.

This was not different in ancient times; one has to go back to the Roman legionary-architect Marcus Vitruvius Pollio (ca.81–15 b.C.) to read about procedures and criteria in architectural design. According Vitruvius, the decision on 'what number to choose?' is given in the proportions of the human body. The numbers of a good design should respect the metric relations between different parts of the body, to be multiplied according the necessity of the project. The use of anthropomorphic proportions and the human body as guiding principle in architectural design was an ancestral tradition adopted from Mesopotamia and Egypt and further used in all West-European cultures, up to the Modulor of Le Corbusier dd.1930 ca. (**Figure 1a-e**).

The same Vitruvius gives indications about the geometry in the architectural project, not directly by speaking about geometric figures, but by explaining the disposition and distribution of each individual quantity. He puts 'ordinatio et quantitas' (in Greek: taxis and posotys) as the first of five conditions, what confirms what was said before on the importance of the 'number'. The second design determinant, the geometry, is included in the 'ordinatio' and 'dispositio' (in Greek: diathesin) meaning the appropriate attention on the three design aspects and image-interpretation, i.e. iconographic, orthographic and scenographic criteria. Furthermore, Vitruvius explains the need for 'eurythmia' (general visual harmony), 'symmetria et analoghia' (harmony and similarity between elements by using a

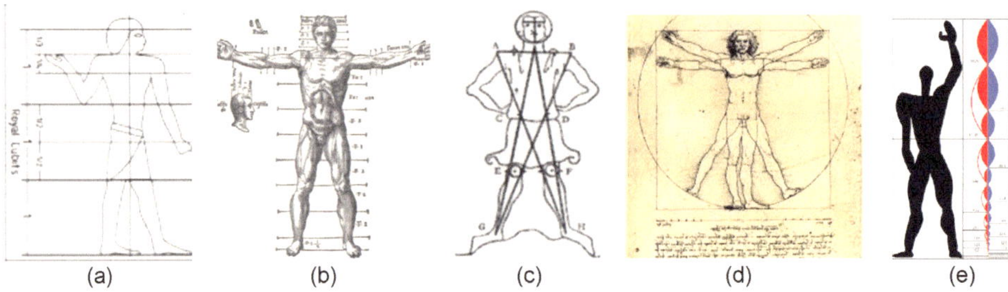

Figure 1.
The proportions of the human body as reference for harmonic design. (a) Ancient Egypt (ca.2000 b.C.), (b) Vitruvius (1st century B.C.; illustration from the Italian edition by F. de Franceschi & J. Criegher in Venice dd. 1567), (c) Villard de Honnecourt (ca. 1235), (d) Leonardo da Vinci, the 'Vitruvian man' (ca.1490), (e) Le Corbusier, the 'Modulor' (ca. 1935).

common 'modulus' for each part of the building, and 'convenientia et distributio (greek: oikonomia)' i.e. equilibrated administration of the available resources and space and, during the execution of the project, a proper division of the costs by calculation ([1], I,1,2). This topics are all well-known fundamentals of architectural theory, and consequently also the fundamentals in architectural design.

As Vitruvius expresses general and universal principles for correct and good architecture, it's evident to find clear applications of his maxims in the design of all medieval buildings, even when there is no written testimony about the spread of his "Ten books of Architecture" in medieval times. The first evident reference to his manual is the drawing of Leonardo's 'Vitruvian Man' (dd. 1490 ca.) as the new standard for artistic proportion and design. This means that the input of Vitruvius's maxims never disappeared, not even after quasi 1500 years.

The one and only authentic written source on medieval design are the 65 pages the Picardian masterbuilder Villard de Honnecourt drew during his travels in Flanders and the Nord of France in the beginning of XIII° century. This sketchbook reveals some of the design techniques of his time, with particular attention for the proportion of building elements and sculptural decoration. Villard does not mention many numbers but shows in his drawings all geometric auxiliary lines and frames to guarantee the correct "euritmia" and "eumetria" in the project. He introduces the aid of a geometric pattern in the design of figures, building plans and elevations, and uses quasi exclusively square or triangular grids. His ideal plan for the "glize desquarie" (squared church) for the Cistercian convent is of particular interest as it represents the model of many West-European medieval churches. Basically, the plan adopts the three naves Latin basilica type with enlarged choir section, flat-ended apse and transept. It is easy to recognize the Vitruvian proportions of a double square module in the central nave axes (longitudinal sequence 6:2:4; transept 4:2:4) and a transversal section in the sequence 1:2:1, with the addition of co-modulated single square lateral naves in both directions. Villard also adds three alternatives for richer gothic choir and apse projects ([6], Taf.28) (**Figure 2a-c**).

In the metric analysis of the St. Francis Church in Assisi [7], we also found the presence of a double grid design i.e. the superposition of a first principal square grid defining the sequence of open spaces for practical use, and a secondary in-between-grid for structural elements (walls and columns) (largely preceding the SAR-design method [7] presented as much innovative in the 1960's).

[7] SAR = Stichting Architecten Research, Netherlands (1964–1990) on the initiative of prof. J. Habraken (T.U.Delft), was working on new housing concepts with large intervention of the future private inhabitants. In stead of the traditional design frames following the axis of the loadbearing structures, they used a doubled design frame with separated spaces for effective utility and for structural elements.

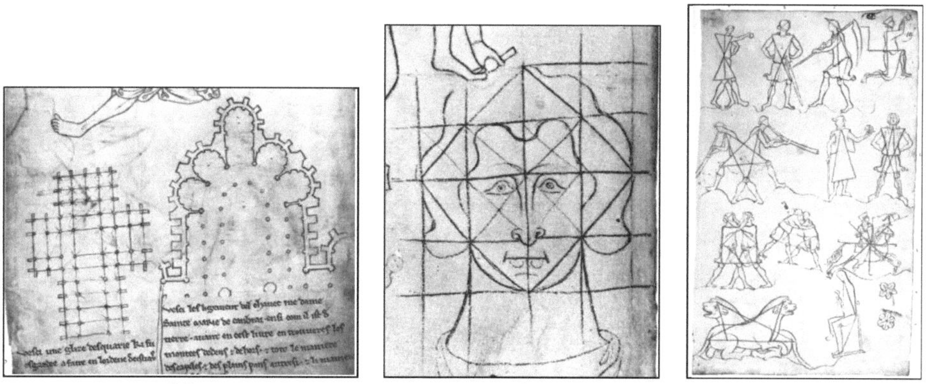

(a) (b) (c)

Figure 2.
Sketches from VILLARD DE HONNECOURT, before 1235 [6]; (a) linear modulated plan of the 'squared church' and ground floor of a large choir with deambulation and radiating chapels; (b) design of a men's head on a squared modulated grid, (c) group of figures, using geometric patterns.

As far as we could check, nor Vitruvius nor Villard use the word *modulus* in their texts, but it is obvious that the 'modulus' must be the key in any design project. This is what Vitruvius intends by combining "ordinatio et quantitas", and what he makes explicit in his definition on "symmetria ... est ex ipsius operis membris conveniens consensus, ex partibusque separatis ad universae figurae speciem ratae partis responses, ..." (symmetry means bringing convenient consensus between all parts of the project, in the separate elements as well as by applying universal forms and figures....) ([6], I, 20–21). He continues by insisting on the use of the dimensions and the proportions within the perfect human body (mutual ratio's between head, chest, arm, palm, foot, finger ecc...) as reference for all design. Villard does not say it by so many words, but presupposes the criterion of mathematical harmonious proportions in the totality of the project as well as in the mutual relation between separated components. He shows it through his graphic analysis and fragmentation of the design steps in his figures (without giving numeric quotes). Also for Villard, symmetry and analogy seem the main criteria.

The choice of the right modulus is the first indicator for the appropriate "ratiocinatione" (= the result of a rational decision-making process) by a most polyvalent-educated architect. In consult with his commissioner, he has to decide about the architectural typology of the building, determined by geometry and size, considering the symbolic capacity of both criteria. In medieval church building, the width of the choir indicated the module for the whole building; the numeric length of the modulus was most symbolic and took normally a number from the Holy Bible or some event related with the purpose of the project. In civil building, the length of the modulus could be of any kind related with the commissioner or the function of the building, as can be seen in the examples below.

Our 'youngest' example dates from ca. 1250 a.D. (Castel del Monte, Andria), but evidently, the modulated design practice did not stop after that period; quite on the contrary! The renaissance architects rediscovered and re-interpreted Vitruvius; and in one way or another, up to the end of the XIXth century all architectural design, working with traditional materials and traditional structures, took profit of the old master's procedures, and 'translated' them in their own contemporary language. But the analysis of the design praxis in the Modern Times is a topic for another study.

4. Selected cases of ancient modulated design

4.1 GIZA (Egypt) Great Pyramid of Cheops, dd. ca. 2570 BC, monumental tomb

Long before the Pythagorean philosophers, the design of this monumental tomb expresses a stupefacient simplicity and coherency in geometry and dimension, fully compatible with the Egyptian vision on the society and the cosmos, witnessing the exceptional culture and knowledge of some 5.000 years ago (**Figure 3a-c**).

The symbolic geometry is obvious: the combination of a square ground floor, symbolizing the flat plane of the earth, and the upwards rising triangular flanks directed versus the celestial globe, with the mummy of the pharaoh and his wife, waiting for rebirth, at the center of the monument. The sides of the square are closely aligned to the four cardinal points. According to some researchers, inside passages and corridors are orientated versus astrologic constellations at the time of building. The King's and Queen's tomb chambers are located in the geometric gravity center of the construction, on emplacement and distances of interior corridors related to the golden mean proportion φ ([9], I, pl.XLIV).

The physical environment of the pyramid, the access way, the sphinx statue and the position of other monumental tombs of pharaoh related people or animals show a well-considered geometric and measured design.

The pyramid's external dimensions witness an equally exceptional design with a selection of most allegorical and harmonious numbering, combining length of sides, diameter, vertical and sloping heights, including also particular prime numbers and the irrational ratio's π and φ (**Figure 3b,c**):

- groundfloor: square (440 x 440) Egyptian Royal Cubits (Erc) of ca.0,524 m = 230,56 m x 230,56 m;

- height (original): 280 Erc = 146,70 m;

- base angle ca. 51,575°, top angle: ca. 76,85°, length slope 220 Erc/cos 51,575° = 354Erc = 185,48 m

- diameter groundfloor ca. 622,25 Erc = 326 m

- ratio (height: side) = 280/440 Erc = 0,636 = ~ 5: 8 (= ~ 0,618 = golden mean) giving origin at the so called 'egyptian triangle', according Plutarch, the most beautiful triangle as it is derived from the equilateral triangle, see **Figure 3b**)

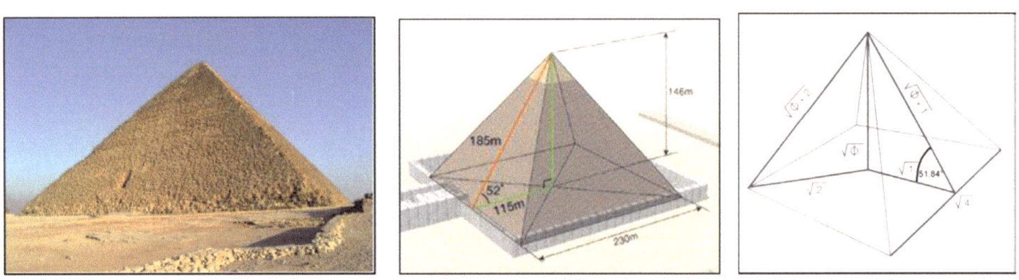

(a) (b) (c)

Figure 3.
Pyramid of pharaoh Cheops at Giza a. general view [8], photo: Giza-legacy.Ch; b + c: Geometries (drawings: Pinterest).

- the ratio (height: ½ side) or the ratio between both cathedes of half of the isosceles profile signs = Erc (280/220) = 1,272727…= $\sqrt{\varphi}$ which results as double irrational by $\sqrt{}$ and φ, and with the exceptional sequence of a repeated number 27 (=3x3x3) after comma, also being easy to draw as it equals the ratio (14:11). This property illustrates the value of number three, being the fundamental (according Plato) of all terrestrial matter.

- ratio (perimeter: height) = 1760/280 Erc = 6,2857 equates to 2π to an accuracy of better than 0.05 percent (corresponding to the approximation of π as 22:7).

- other surprising arithmetic ratios and golden- mean- related dimensions are signed in **Figure 3c**.

The integer dimensions (440 and 280 Erc) as well as the irrational ratio's 5:8 and 2π = 2(44:14); (note the numeric resemblances between 44 $\leftrightarrow$ 440 and 2(14) $\leftrightarrow$ 280) also include (unknown) allegorical messages. [8]; [Wikipedia].

4.2 ROMA, pantheon - founded ca. 23 b.C.; present structure dd. ca. 125 a.C. - temple

This ancient Roman temple, dedicated to all God's, is a most interesting monument for many reasons, a.o. for his unusual design showing a multilayered intangible content(**Figure 4a-d**). The building was founded as a rectangular temple about

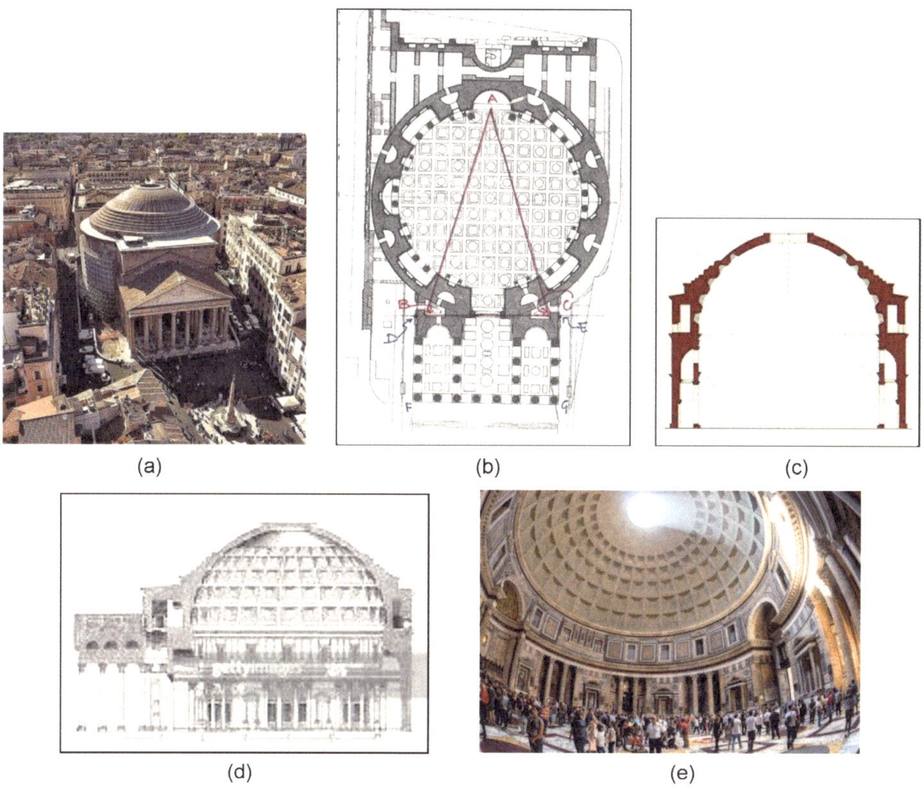

Figure 4.
Roma, pantheon: (a) aerial view (photo Pinterest); (b) ground floor of rotunda and so-called portico (Wikimedia & author); (c) transversal section (Pinterest); (d) longitudinal section (Getty images); (e) general view rotunda inside (Wikimedia).

27 b.C. by Marco Vipsanio Agrippa (ca. 63–12 b.C.), rebuilt two times, in his present form by the Emperor Hadrian in ca.125 a.C.. The monument shows two distinct parts: the mayor part or 'Rotunda' with cupola, and a squared classic temple, today serving as entrance area. The overall image shows two buildings, with two separated spatial identities, two formats and two functions.

The Rotunda represents the allegoric bricked envelop of an impressive regular globe. The horizontal diametrical plane of the globe divides the interior in two equally high volumes: the upper part shows a half-sphere dome, structured with cassettes and at the top an open oculus, the only entrance of natural light; the lower half is a cylindrical volume, for his part divided, according the golden mean ratio, in a lower section including a sequence of niches, apses and columns, and a upper dome-tambour section elaborated with squared cornice patterns. The cylindrical walls of the lower half of the interior are richly decorated by different kind of materials and forms. The Rotunda is the evident metaphor for the cosmos with the dome as the celestial half-round and the cylindrical lower space representing the terrestrial world, everything dominated by the central oculus representing the Supreme Divinity, generating life and dynamism through the zenithal light entering from the oculus .

The second part, the so-called portico, minor but nevertheless substantial part of the design, seems conform with one of the traditional Greek-temple-inspired models described in Vitruvius manual ([1], I, III, 284). However, the design does not follow this models too much; on the contrary, it is much closer with the most ancient tripartite Etruscan-Italian temple as described below. This deviant design from the conventional temple-pattern, seems an explicit demonstration of the own native Etruscan and Italian temple-building origins, different from those imported by the first Greek colonists in the 3th and 2th century b.C. It is as if Agrippa Vipsanio, commissioner of the first temple and son in law of the first roman emperor Antony-Augustus (with divine status), or one of his successors involved in the reconstruction of the temple during the 1st and 2d century a.C., liked to stress the native identity of the italic people, denying the fact of roman sacral architecture being indebted to the Greek (although this clearly appears from the first temples e.g. in Sicily and Paestum, or from the great basilica's built on the Fori Romani). The squared portico part has a net floor area of circa half that of the Rotunda, which is much too large for being only 'portico' (or 'pronaos' as it is wrongly called in some literature). The approximate ground floor surface ratio 1:2, and the change from the rectangular versus the circular format, rather seems a conscious combination of old and new, an indication on the start of a new era for Rome, reminding the remote origins of the first Sannite (origin of the temple's founder Agrippa Vip-sanio?), Tuscan or other settlers in Central Italy, and Rome's passage as the capital of a new Mediterranean and European empire. The creation of the square antichambre-like temple before entering the incomparable grandeur of the Rotunda reinforces the expressivity of this last one. Both united entities are a rare example of architectural design as a political statement, materializing history and social order.

- The geometry of the double monument

The circular Rotunda: as easily deductible from (**Figure 4a-e**), the geometry of the Rotunda is a most regular compilation of circles, squares and even a equilateral triangle (transversal section) in bi-dimensional and tridimensional edition, all of them having their specific semantic content. The rhythmic alternation of the tripartite 'negative' savings and 'positive' porches on the lower cylindrical ring, and the polychrome with the changing incidence of natural light from the oculus at the top, creates a great sense of wealth and dynamism. The frequency of the number eight in the design, by quantity (e.g. 8 mayor interior savings and 8 minor exterior ones in the perimeter wall) and by dimension (e.g. the overall inside diameter of 146 roman feet – see below), as well as the ancestral semantic of this number (being the first cubic number as $8 = 2^3$) do believe that eight (and his composing factors 2–3-4) is the numeric modulus for the design of the complete building (to be completed with the number 5 as we'll see further). The modulus should be found also in the net width of the opening of niches and apses, but the necessary metric information to prove such assumption, was not available. Apart from the 8 mayor niches/apses, the interior parietal composition includes 8 jutting out porches; both artifacts create a imaginary cylindrical space-filling web of 8 + 8 = 16 isosceles spiked volumes, pointing to the central vertical axis connecting nadir and zenith of the as imaginary interior sphere.

The squared entrance-temple is designed according the archetype of the three-cellae open Tuscan temple surrounded by columns and divided by two intermediate rows of three columns. The net inside width of the three cellae-areas is traced according the golden mean ratio. The two lateral cells have a small apse at the end for some simulacrum, the wider middle area covers the common central longitudinal axis with the double entrance door of the Rotunda, and is orientated versus the head apse of this Rotunda with the statue or the seat of the emperor. The entrance-temple is surrounded by eight columns at the front and three columns at each side with a short in- between piece of wall connecting with the Rotunda. The temple front looks similar with the Vitruvian models but other research is needed to identify all metric differences. After conversion of the Pantheon ensemble in a Christian church in 7th century, specially this entrance part and the in-between connecting structure suffered various amputations of materials and additions with demolition afterwards of crowning towers, but the present image should approach the authentic one.

- The arithmetic at the double monument

The choice of the numbers, i.e. the dimensions and the quantities of the decorative elements (niches, apses, columns, porches, cornices, marble paneling), as well as their spatial distribution are additional indicators for the intangible messages. This chapter cannot enter into the detailed design aspects of this elements, but one can safely conclude that the Pantheon ensemble is probably the very first example of a fully integrated Gesamtkunstwerk based on the ancestral symbolic geometry and numbering; this last by using the very basic integer quantities 1–2–3-4-5-(1,618= φ), and their derives 6–8-10 and other. We illustrate this by the only three beginning and most decisive choices of the design process.

- Three decisive choices at the start of the design process

1. The very first step concerns the choice on the architectural typology and form, based on the functional requirements and the symbolic content, mentioned before. This resulted in the option to create an innovative cosmic sphere imitation in a format which was never done before; to be connected with a reminder of the presumed architectural origins of Rome, i.e. the Tuscan temple. The innovation is proved by the technical audacity and capacity to build a dome structure signing the incredible span of ca. 43,40 m or 146 rf, which is the largest span ever in building from antiquity up to mid XIX° century! The 146rf dimension is certainly not arbitrary; the radius of 73rf = 1 + 8 + (8x8) might refer to the number 8 = 2x2x2 as the probable numeric modulus for the entire

composition. Unfortunately, we do not know what symbolic content Agrippa or Hadrian connected with the number 8 modulus.

It's interesting to notice also the open oculus diameter of ca. 8,35 m = ~ 28rf which signs ca. 28:146 = ~ 1:5 ratio to the central diametric plane of the sphere, and a reference to the semantic of number five, the golden mean ratio's (including $\sqrt{5}$) and the pentagonal symmetries in other parts of the building.

2. The second decision concerns the likewise exceptional thickness of the external wall of ca. 5,90 m = ~ 20,00 rf (=2x2x5). Considering the presence of the deep niche savings, the minimum thickness of the external bricked shell is reduced at ca. 2,36 m ~ 8,00 rf (including the todays disappeared external marble wall cladding), which seems comparable with other antique buildings. This quote is not the result of any structural calculation (although technical experience must have been involved), but of an exclusive geometric property. Indeed, the external diameter signs ca. 55,20 m or 186,50 rf, and his ratio to the inside 146rf diameter equals 186/146 = 1,274 = ~ 1,2727 = $\sqrt{1,618}$ or the square root of the golden mean proportion (with connotation of sequence after comma 27 = 3x3x3 similar with Cheops ratio). This means that the length of both diameters (external versus internal) are part of a Diophantic triangle progression[8], referring once again to the golden mean proportion φ = 1,618 ([10], Ù 49). Such arithmetic property only occurs in exceptional cases, and is highly appreciated as the irrational square root as well as the irrational φ value results a double infinity reference, most symbolic for the cosmos, for the divine status of the emperor Augustus and for the future of the Roman empire [9].

3. The third decision regards the dimensions of the entrance-temple, signing a double squared ground floor of ca. (31x17,5)m = ~(104x59)rf or a normal length to width ratio of approximate 2:1. Archeological excavations at the end of 19th century have proved the presence of more steps and a normal stylobate space to get on the columned entrance area; this means that the squaring could have been slightly different from today. More important however is the geometric connection between the Rotunda and this open temple structure. After searching and calculating possibilities, the author found out that the net frontal width BC (free passage between the side walls) of the portico (ca. 31,0 m) is given by the base of the 'golden' or 'sublime' triangle ABC inscribed in the Rotunda, with the top A in the center of the head-apse and the base along the inner line of the double (formerly bronze) entrance door of the Rotunda. (**Figure 4b**). The 'golden triangle' is found in the spikes of regular pentagons and decagons, and is a isosceles triangle such that the ratio of the hypotenuse to base is equal to the golden mean. The calculated hypotenuse signs 48,45 m (diameter from apse to entrance door) ÷ cos. 18° (half top angle of 36°) = 50,946 and 50,946 ÷ 31,0 (measured width of the portico's front) = 1,643 or, within normal tolerances, identic to the golden mean ratio 1,618.

[8] Called after the Greek mathematician Diophantes of Alexandria (ca. III° century AD). The property of a Diophantic triangle means that the length of his sides can be expressed as part of a geometric sequence with a reason (i.e. the constant proportion between two successive numbers of the sequence) that equals 1,272 or $\sqrt{\varphi} = \sqrt{1,618}$.

[9] Many of the antique geometric and arithmetic symbols have been adopted by early Christianity; the first Church Fathers and Scholastics christianized the ancestral pagan semantics, converting and adapting them into biblical and christian concepts.

The next design step draws a double square, sided ½ 31 m, at the left and the right of the central longitudinal axis to create the overall entrance-temple area DEFG. Further on, each of both composing squares get divided according golden mean ratio with the width of the lateral bay (with end-apse) as the ratios mayor and half of the central area as the ratios minor. The joining of both minors results in the central area along the common longitudinal axis of the ensemble. By this procedure, the entrance temple and the Rotunda get physical (through geometry and numbering) and spiritual (through various semantic) most intimately connected.

We notice, once again, the application of the golden mean ratio. The frequency of this ratio in so many design procedures indicates his particularly powerful allegoric meaning as indicator of cosmic harmony in life and society and of rebirth and infinity of man.

4.2.1 Conclusion

This concise analysis discovered several unexpected qualities of the Pantheon ensemble. They are not simple architectural 'curiosities' but existential part of the building's identity. The multilayered image of this 2000 year's old ensemble shows the resilience of simple but conscious design and his timeless capacity for tangible and intangible communication.

4.3 AACHEN (Germany), Carolingian Imperial chapel, start 795 a.D., dedication 803 a.D.; (extended with gothic choir and several side chapels from ca. 1350 onwards)

The chapel of the imperial palace of Charlemagne is another example of the impact of the geometry and the arithmetic in the dialog between the building and his observer. The commissioner is the first West- European emperor Charlemagne (742-814 a.D.) after the fall of Rome. For the chapel of his palace, he chooses the model of what he might have seen on his conquests (e.g. S. Vitale at Ravenna dated ca. 530 a.D.) and what linked him with his illustrious predecessors. He adopts, for the first time applied on this scale in the northern-of-the-Alps countries, similar innovative design which includes a lot of Christian and imperial symbolisms and allegories. We refer to the architectural history books for all details – e.g. see [11]. Our limited notes mark the most evident design characteristics employed as tangible instruments in the communication of intangible contents .

- The geometry (unit of 1 cubit = 0,4281 m [12])

The plan: although the external image of the Carolingian building looks almost circular, the fundamental plan concept is of a squared design, i.e. a central octagon, which is the result of two superposed identic squares of which one is rotated over 45°. The central octagonal area is surrounded by a ring of eight squared chapels connected by eight triangular interspaces, generating the hexadecagonal external envelop. The central area is connected with the ring area through eight arched passages, covering quasi the full length of the octagon's side (**Figure 5b and e**).

The vertical section shows two concentric volumes; the central octagonal higher one, open up to the top, and the surrounding hexadecagonal ring of two levels: the ground floor spanned by cross ribbed vaults and a upper gallery similar to the Romanesque matroneum, looking into the central octagonal space through a double

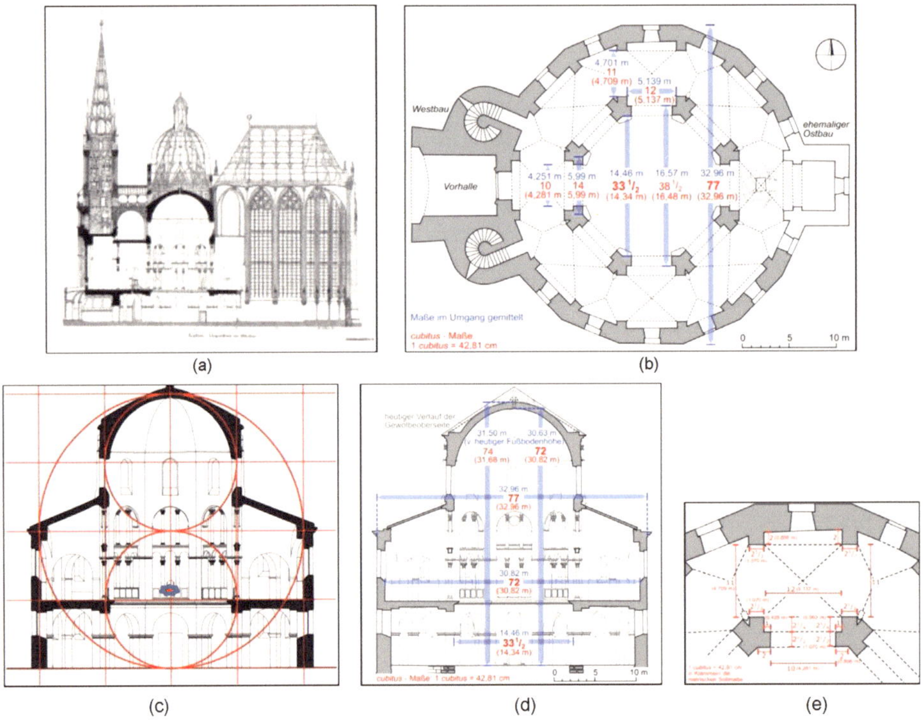

Figure 5.
(a) Aachen, our Lady's cathedral, longitudinal section with the carolingian chapel at the west half and the gothic choir (14th -15th century) at the east half of the ensemble. The tower above the west entrance and the raised dome above the octoganal Centre date from 18th & 19th century (drawing Kunstdenkmäler Rheinland, 1916). (b) Aachen, our Lady's cathedral, ground floor of the end-8th century Kaiserkapelle (drawing from [12]). (c) squared proportions of different sections of the building and indication (star) of the emperor's chair at the center of the lower circle (d): Arithmetic modulation and ratio's in the octagon; (e): Mutual proportions in the squared ring-chapels / drawings from [12].

superposed tripartite arched opening, divided by two columns. With the outside windows at the top of the octagon, the central area gives the impression of a four-level structure, where the physical evidence of the surrounding ring-volume counts only two floors (**Figure 5c**).

The center of the chapel forms a regular octagonal prism; the overall circumscribing volume marks a regular virtual cube, with the sides equal to the hexadecagonal's diameter, inside of ca. 30,82 m = 72 cubits, and outside of 32,96 m = 77 cubits. The prism's inside height from floor to the top also signs 72 cubits. The outside top-cornice of the hexadecagonal ring signs the precise horizontal middle plane (height 36 cubits) of the chapel, separating virtually the ground floor with the upper gallery of the imaginary 'terrestrial world' from the single globe on the top, inscribed in the dome and tambour volume, the imaginary space with the presence of the one God. Some parallel can be made with the domed upper half (with oculus) of the Roman Pantheon, apart from the octagonal perimeter versus the roman circular one. The sequence of spaces in the lower surrounding ring volume applies a most regular and symmetric geometry with congruent cubic volumes alternated by triangular spikes on both levels of the ring space (**Figure 5c**).

The amalgamation of the three most fundamental geometries: the square (part of the octagon), the circle (inside and outside virtual volumes, plus the staircase towers flanking the entrance), and the triangle (the eight spikes inside the hexadecagonal ring-space) generates an exquisite allegoric ensemble, open to all kind of profane and religious interpretations. In addition to the predominant centralizing plan-concept, the entrance extension with two massive towers in the west generates a additional east–west longitudinal axis (different from the Ravenna example[10]) with the emperor's marble coronation armchair at the center of the upper squared floor section, looking east towards the central golden altar. Although the setting of the armchair might have been different at Charles's time, it signs the beginning of a later developed Romanesque tradition to build a special imperial room on top of the west-entrance of more double oriented churches, as one can still admire in some large Meuse and Rhine churches (e.g. St. Servaes at Maastricht, Netherlands)[11] .

The hexadecagonal chapel was a integrated part of a squared parceled design covering the complete imperial palace site, including an open forecourt to the chapel, two small and one larger basilica's for the emperor's activities as well as several residential buildings. On top of the sophisticated chapel geometry and the squared environmental design, there should have been involved also astronomic considerations related with the orientation of the single buildings and, concerning the chapel, with the incidence of natural light, the original emplacement of the altar, the location of the emperor's chair, and other particularities, but this seems not enough studied yet.

- A rich arithmetic reading (**Figure 5b-e**)

The dimensions and quantities in the Imperial Chapel reveal the intended communication by specific dimensions. It starts with the choice for the octagonal plan, derived from the square, and including the semantics of the number four, eight, sixteen, and the multiplication factor two, appearing frequently in the building. The single entrance with the two flanking towers maybe symbolizes the double nature of his commissioner as worldly and religious sovereign, or also refers to the two columns crowning the entrance of King Solomon's temple (1Kings,7,15–22) indicating imperial dynamism and power. The overall design might have been guided by Alcuinus of York (732–804), chief philosopher and theologian at Charlemagne's court, as all mayor dimensions, after conversion into the then local cubitus (= ~ yard of ca.41,81 cm), result symbolic for biblical quotes. In this perspective, the design modulation does not appear that much a physical condition for related quotes, but a mental and metaphysical set of numbers taken from the Holy

[10] It's interesting to compare the *Aachen* design with the older S.Vitale at *Ravenna* dd.525–547. This last one applies the octagonal plan for the central area as well as for the surrounding ring-volume. The passage between the central and the perimeter ring space in Ravenna is made through tripartite bowed niches and the exterior wall has a different windows distribution, which result in a more open and better illuminated interior space. There is a double entrance, passing through a full-width narthex, deviating from the central symmetry axis. The overall design includes a rich allegoric content by geometry and arithmetic.

[11] One has to realize the different character of Charlemagne's palace chapel, deviating substantially from the current typologies for private chapels in aristocratic palaces or castles of the early Middle Ages. Normally, such chapels were built as two strictly separated entities: a upper volume at the exclusive use by the noble men, and a lower volume (without any direct connection with the upper one) for the use by the lower social classes, or in many cases also as memorial chapel for the family tombs of the seigniors. Such physical separation was done for evident security reasons, but also as confirmation of social hierarchy. Charlemagne introduced here a different typology: on one hand a octagonal sanctuary in his double identity exclusively at his personal service with double floor level and double horizontal areas i.e. the central octagon and the surrounding chapels, all in one volume, visually connected and also physically by two stairs; and on the other hand two traditional rectangular small basilica's, built at the outside left and right of the hexadecagon, both at the service of the palace staff.. .

Bible and the ancestral arithmetic symbolisms passed down from the Pythagorean philosophers. Nevertheless, there is great logic and uniformity in the geometric figures as well as in the choice of integer and simple numbering for the design of length, width and height[12] .

This logic is found in the harmonic multiples of small (metric) quantities, and simple mathematical sequences (e.g. the arched openings connecting the octagon and the hexadecagon sign 10–12-14 cub. (**Figure 5e**). As said, there is preference for prime numbers and biblical numbers, many of them borrowed from the Apocalypse and the there description of the celestial Jerusalem (Apo.21, 9). This Jerusalem is a squared city (the octagon signs the superposition of the terrestrial and the celestial Jerusalem), and was equally long, large and high, with a wall of 144 (=12x12) cub. Also the virtual cube of the chapel volume is equally high and large (inside 72 cub., outside 77 cub.[13]); the longitudinal side of the squares in the ring-volume signs 12 cub. and the transversal opening signs 11 cub.. The ratio 12:11 has his own semantic with 12 referring to the 12 apostles of Christ or to his presumed 33 year's life time on earth (to read as 12 = 2x(3 + 3), meaning the double nature of Christ (divine and human) living during 33 years; and the 11-quote indicates men's imperfection while Christ's perfection is signed by 12; to be seen also as 11=1/3 x 33 of Christ's lifetime and the metaphysical passage (12=4x3) from a squared central space into the triangular inserted spaces of the hexadecagonal ring.

The passages from the octagonal area into the surrounding area sign an opening of10 cub.(with 10 being the 'perfect' quantity). In most medieval churches, the architectural modulus is indicated by the passage opening between apse and choir; this brings us to conclude that also in this case, the metric modulus has been ten cubits (= ~ 4,18 m), eventually in combination with a secondary 12 cub. Quote as a recall at the left and right of the passage opening (**Figure 5e**). The octagon's diameter of 33 ½ cub. is another reference to Christ's life time on earth (**Figure 5d**); the diagonal of the hexadecagonal and the altimetry design, signing the same quote of cub. 72 = 2x3x12 = 3(2x3x4); the external diameters of the octagon (38 ½ cub.) and the hexadecagonal (77 cub.) sign the evident proportion 1:2. Also the number 77 should not be fortuitous because, according the calculations by S. Augustine of Hippo (354–430), (based on the Old Testament history data), the number of generations from Adam to Christ counts 77! [12]. The lecture of this building's dimensions seems a florilège of biblical quotes, a very typical phenomenon in medieval church design .

4.4 CHARTRES (France), Our Lady's Cathedral : start rebuilding 1194 (after fire damage), dedication 1260

The design of this French gothic key monument followed the then current design criteria: a ground floor ad quadratum according the église carrée of Villard (**Figure 2a**), and an elevation ad triangulum. Although the available metric documentation [13–17]; [18], p. 84, is not uniform, the evidence of a coherent modulated design is undeniable, including the inevitable smaller or bigger differences

[12] The analysis of the measured drawings from [12] proves once again that ancient modulation normally refers to the free passable space in between structural elements as walls and piers. Quite rarely also the thicknesses of this elements are included in the modulated design or are part of the allegoric dimensions (except from the Diophantic geometry as it is the case for the Pantheon's exterior wall).

[13] The outside height of the dome shell actually signs only 74 cub. as the roof structure as well as the ground floor quote has been changed in a later period; the initial height should have been also 77 cub. [12].

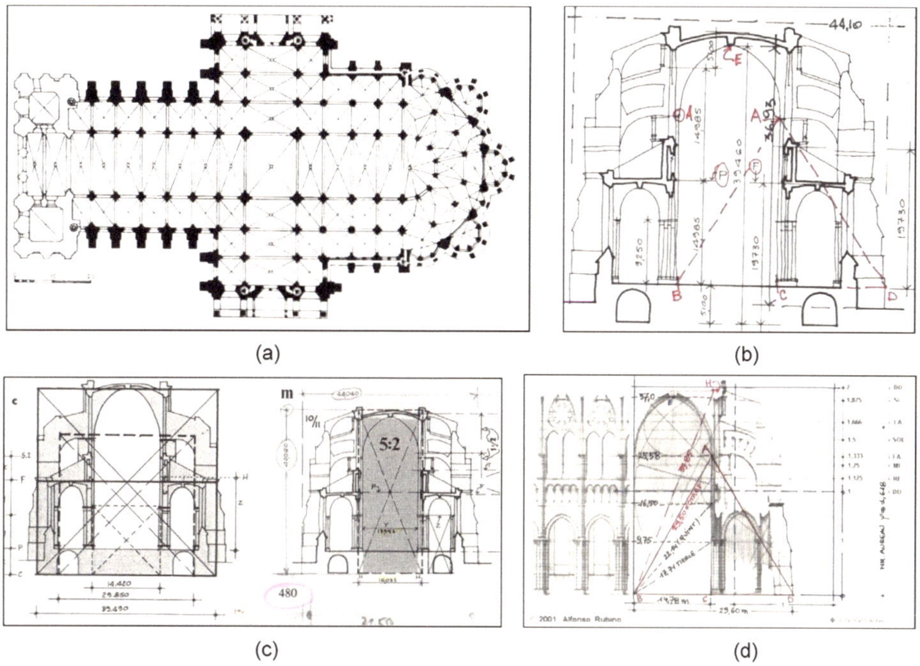

Figure 6.
Chartres, Our Lady's Cathedral. (a): Plan [13]. (b) Transversal section [16] and author. (c) Ratio's transversal section (drawing from [16]). (d) Hypothetical altimetry design according musical canon [Wikicommons + A. Rubino + author].

and deviations, caused by ca. 100 years construction activity by different master-building teams (**Figure 6a**). Similar to what we'll see in Assisi, also in Chartres there is the very first question about what might have been the metric standard applied on the building site because of the confusion in the historic terminology and the uncertainty on the real length unit in practice at that time: or the French "coudée" or cubit of 0,5236 m, or the ancient "pied du Roy" of 0,3236 m or maybe still another local standard .

As said, the metric modulus of the medieval church was given by the net passage opening between the apse and the choir or the choir and the nave, in this case signing **14,78 m** (and not the 16,40 m distance between the center-to-center columns axis's, as kept on in most publications)! Our reference publications propose different 'fabricated' measure units, quite close to the "pied du Roy", as e.g. 0,369 m in [14] or 0,333 m in [18], partly inspired by the Christian semantics associated with number three and his multiples. However, none of those are coherent nor with any historic evidence. By reasoning back to front, comparing with other medieval churches, and inspired by the case of the St. Francis Church in Assisi, it seems probable and plausible that the Chartres's bishop in charge, out of respect to the Roman Pope, absolute head of the Christian church, and maybe also out of piety tradition, imposed to practice the ancient roman foot standard of **0,296 m**, applied in Rome and in many Christian church buildings. This same standard was called in France also "pied de Cluny" for the evident reason that the Cluny convent and the Cistercian monks, since late 10th century, were the most active church builders in Europe. Converting on this base, we obtain a most sense full and most Christian symbolic modulus of 14,78 m: 0,296 = **50rf** (roman foot). The acceptability of this "pied de Rome" unit is confirmed in the converting of some representative dimensions of the building (always considering the only free passable spaces).

However, for want of a recent complete measured drawing, we only note following data (**Figure 6a**):

- Architectural modulus: 14,78 m → 50rf = 2x5x5
- Transversal width of choir-bay (estimated) 6,25 m →21rf = 3x7
- The overall design grid in the choir and the nave signs a sequence of oblong rectangles and squares with slightly variable dimensions, apparently depending on the section of the columns and alternative ratio's in the tracing of a virtual geometric grid. The grid indicates the net squared areas of the quadrangles, separated by the construction strips, corresponding with the width of the vault-arches. The quadrangles can include different semantics such as 'golden rectangles' (φ related) or 'dynamic rectangles' ($\sqrt{}$ related).

  According our scaled drawings, the nave and choir rectangles sign theoretically (50x21)rf, the aisles (25x21)rf and the crossing (50x42)rf. All quadrangles are semantically considered equivalent with regular squares.
- Global inside length (from the entrance porte royal to the closing wall of the apse): 130,20 m → = 440 rf or 400rf of church passage and 40 rf of entrance zone between the towers;
- Global inside width of nave and two aisles: 32,40 m → = 109,46 rf or 100rf of passable church area and 9,46 rf for two rows of pillars construction zone;
- Global height inside the choir (from first outside stair- threshold of the entrance, i.e. the public space area, to the underside of the vault-keystone: 36,93 m → =125 rf = (5x5x5)rf (**Figure 6b**)
- Global height from crypt-floor to nave vault: 39,46 m → 133,33 rf (to read as separated ciphers, with double reference to the 33 years Christ lived on earth)
- As shown in (**Figure 6b**) there is no doubt about the elevation design ad triangulum with the modulus BC equal the half-side of the equilateral triangle with top A at quote 25,58 m (= 29,50x1/2 $\sqrt{3}$) , indicating also the impost of the nave-vaults arches and the bottom of the first outside flying buttress.
- The top E signs the intersection of the two vault curves 'at the fifth point', at the height from outside threshold of 37 m in a ratio versus the width of the nave (= modulus) of 14,78 m:37 m = 1: 2,5. The modulus signs half of the length of the diagonal AB (=side equilateral triangle).
- The external transversal section of the nave(including the crypt areas) describes a quasi-squared plane (height 40 m x large 44 m) = 10:11, which signs the already noticed semantics. The capitals of the triforium corridor (quote 19,73 m) sign the geometric middle of the full inside elevation; the overall transversal section on the nave forms a 5:2 rectangle (**Figure 6b and c**).

In this limited list, we see more references to number five and his multiples, including the number 100 and 400 as reference to the exclusive hecatompedon quantity (i.e. 100 = 10x10 what means a more than perfect number) and a supplementary indicator for the 50rf as present modulus.

To conclude, we have to mention the interesting hypothesis on the relation between architectural geometry canons and the music harmony canons, which seems to find confirmation in the elevation of Chartres's nave (**Figure 6d**). Just as good architecture should bring order in a chaotic space, also harmonic music is as a cosmos imposed upon chaos [19], 251ff. Considering the side-length of the equilateral triangle (= double modulus) as an octave interval, one connects some significant altimetry quotes or virtual lengths from the opposite floor-bottom, with the tierce (lower capitals of central piers), the quart (keystone vault aisles), the quint (passageway triforium), sext or sept (lower threshold nave windows), first octave (impost vault-arch nave), second octave (~ keystone vault-arch nave). This needs further study before being confirmed.

4.5 ASSISI, S. Francis Basilica; start 1228 - dedication 1253 (extended in later decades)

The S. Francis Basilica at Assisi was built to worship the sepulcher of Francis of Assisi (1182–1226), founder of the mendicant Order of the Minorite Fathers. The work started March 1228, but after two years, mid 1230, the project mission was extended with a second assignment i.e. becoming the representative mother church of the new religious Order. Integrating this new function in the same, still under construction, sepulchral and pilgrimage church, did not seem possible and for want of space on the same site, it was decided to build a fully separated second church on top of the first one, keeping the same external wall's perimeter and a similar inside spatial distribution. Such audacious project got realized after reinforcing the already built exterior walls and after inserting a new type of cross-rib vaulting for the lower church. This phased realization resulted in the actually superposed double-church, characterized not only by two different functions (a Lower devotional Church for S. Francis's tomb and a Upper Church for the Fathers Convent's services), but also by a double architectural and artistic identity: a first late-Romanesque Umbrian Lower Church and a second early-Gothic European Upper Church.

The design of the Lower Church (excluding later extensions) adopted the traditional middle sized Umbrian single-nave and single level church model. The geometry and the arithmetic's on dimensions and quantities of the Lower Church integrated the ancient Pythagorean traditions, modifying them according Christian semantics with tangible imitation of some iconic Christian churches of that time i.e. the S. Peters Basilica of Rome and, as it was built in full crusaders period, also the S. Sepulcher of Jerusalem, and an allegoric record of the since long demolished biblical Temple of King Solomon on Mount Moriah in Jerusalem. This multifaceted mission was realized in a multilayered design, building a quite old-fashioned Romanesque Lower Church and an innovative Upper Church, this last one in the new transalpine (French) gothic style and structure, the very first application of this new architecture in Italy. It's interesting to notice the remarkable coordination in the design of dimensions, forms and structures within two architecturally and structurally so different buildings. Under this prospective, the S. Francis double church is a interesting example of design resilience and flexibility avant la lettre [7, 20, 21].

- The geometry

Notwithstanding the two-phased origin of the S. Francis Basilica[14], the geometry and the arithmetic follow the ancestral symbolisms of figures and numbers, independent from the clearly different stylistic character of each church and the references in orientation, distribution and architectural forms to the three mentioned iconic Christian churches .The plan-geometry of the S. Francis Basilica is a evident ad quadratum design, persisting in the early XIII° century four bays Romanesque Lower church, continued in the mid XIII° century Upper Church and the XIV°-XVII° side chapels. The modulus is given by the width (11,84 m = 40 roman feet of 0,296 m) of the passage in between the apse and the crossing. As a one-nave church, the initial three nave-bays sign a perfect (40x40)rf square, separated from each other by a narrow strip for the transversal arches. The crossing signs (43x43) rf in order to realize an inside area with hecatompedon overall length from the XIII°century entrance door to the end of the apse. The bell-tower adopts the same modulation, applied however on the outside perimeter. The design of the later additions and modifications (the superposed second church, the east entrance-transept of the lower church and seven of the twelve lateral chapels) followed similar squared design. The lateral chapels however applied a reduced modulus of 23 or 24 roman feet (**Figure 7b**). The geometry of the two transept sections (added in a second campaign) are two 'golden rectangles', applying the same 40 rf modulus in longitudinal direction and defining the transversal width according the golden mean proportion (40 rf x 0,618 = 24,72 rf = ~ 7,32 m). The overall analysis of the modulation permitted the author to discover some revisions and changes in the design, realized in the course of building, and to identify the probable chronology of each section of this medieval project [7].

The vertical section of the Lower Church signs a surprising ad circulum design, i.e. each bay includes a regular sphere with a diameter equal to the 40 rf modulus. The nadir of each sphere does not coincide with the pavement level, what should be the normal design, but with the quote of the Saint's sarcophagus, below the pavement. The sarcophagus is located at the base of a virtual sphere, inscribed in the crossing of the nave and the transept. The same sarcophagus marks also the starting point for an imaginary vertical axis, passing the middle of the mayor altar, rising to the zenith, as to indicate that the remains of S. Francis, a unique relic treasure with exceptional thaumaturgy capacities, is the best go-between to resolve men's problems. To stress this capacity, the difference between the extrados and the intrados perimeter of the circular transversal ribs delimitating the crossing, have been calculated according to a 1/3 reduced width of the diophantic proportion (similar to the Pantheon's enveloping wall) indicating the double irrational and golden mean reference of $\sqrt{\varphi}$ (**Figure 7d and f** [20, 21]. Both churches also include a symbolic parcours on rebirth, indicated by the geometric path from the Lower Church virtually ascending through both apses and the Upper Church [23], p. 117.

As the initial design provided with two identic churches, both ground floors are similar apart from the nave's interior width, which is little larger in the Upper Church as the exterior wall thickness of the Upper Church is only half of those of the Lower Church. However, the opening between apse and crossing also in the Upper Church keeps strict on the architectural modulus of 40 rf, notwithstanding

[14] The denomination as 'Basilica' is rather misleading as the church is not conform with the historic basilica building type; it is a exceptional honorary title, awarded in 1756 in occasion of the 500th years anniversary of the dedication of the double church, to give the S. Francis church the same ecclesiastic privileges as the ancient Christian Basilicas in Rome.

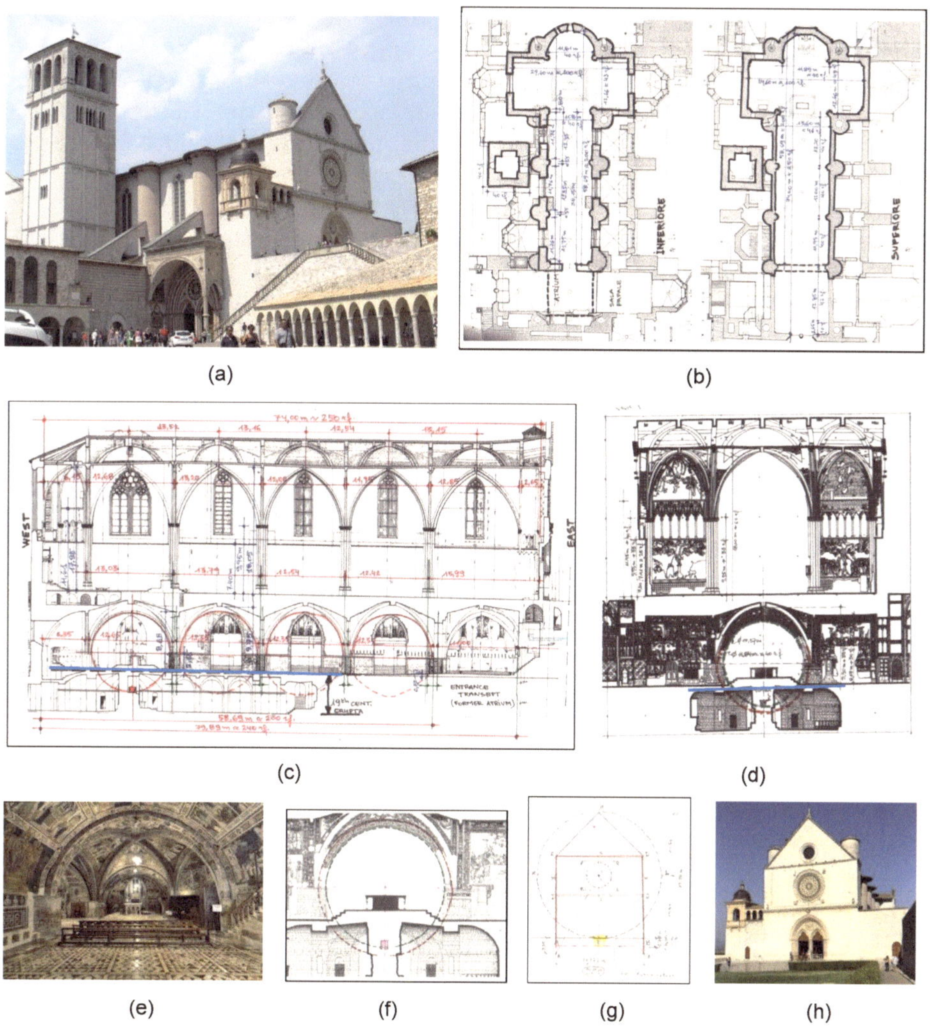

Figure 7.
Double St. Francis basilica, Assisi. (a) General view from south-east [Wikidata]. (b) Squared ground floors lower basilica (left) and upper basilica (right); the black delineation indicates the mid-thirteenth initial three-bays church, the gray volumes the later additions [22] & author. Francis basilica, Assisi. (c) longitudinal section over both churches, extended in XIX° century with crypt. (d) transversal section over the transept (drawings [22] and author). S. Francis basilica, Assisi. (e) Lower church, view on the main altar, versus west [Wikimedia]. (f) Transversal section over main altar lower church and XIX° century crypt with location of sarcophagus in the nadir of reduced diophantic sphere (FOART measuring & author). (g, h) Geometric design of the east façade upper church (drawing and photo: Author).

the different geometry of the apse (circular in the Lower Church and decagonal in the Upper Church). The stylistic differences in ornamental design (gothic versus romanesque) is also very evident as this emerges e.g. in the massive trilobate wall-piers of the Lower Church in contrast with the five-lobate clustered piers in the Upper Church. We also noticed several metric design irregularities, e.g. the vertical axis' of the Lower and Upper Church wall-piers are not well centralized, nor the length of the bays and squared plan-grid of the Upper Church are very regular. This are obvious indications for the separate design of both churches, spread over more years and more chief-master builders; maybe also connected with the medieval concept on metric tolerance's and the not so perfect measuring instruments. However, the visual impact of this metric irregularities is negligible as they get disguised by the full polychromic decoration of all walls and vaults.

The vertical geometry and the wall's elevation of the Upper Church is very different from the Lower Church. As said, the Upper Church design expresses the introduction of the gothic design canons, inspired by the gothic examples in the French and English Normandy and the Scholastic church fathers. The design implies a predominant ad triangulum geometry and the application of the different types of pointed arches as indicated by Villard de Honnecourt [6].

- The arithmetic numbering (conversion of metric unit: Roman foot = 0,296 m)

The numeric quantities in both churches reflect a large collection out of the mentioned sources from pagan antiquity, the Holy Bible, the Christian Scholastics, and the practical need for simple and rounded numbers. It starts with the choice of the horizontal as well as the vertical modulus in both churches, equal **40 rf** (a number appearing more than 100 times in the Holy Bible and also the modulus of King Solomon's Temple); the overal inside length (from apse to entrance) of the sepulchral church signs 200 rf (=5x40), the inside width of the transept signs 100 rf, i.e. a single and a double hecatompedon (the length of the mayor antique Greek temple); the inside length of the extended Upper Church signs 250 rf; the free height in the Lower Church from the pavement to the top of the transversal arch signs 30 rf and 33 rf to the vaults keystone signs; the piers' impost in the Upper Church signs also 33 rf; the height of the vaults keystone signs 60 rf or a 1:2 proportion regarding the Lower Church height. In the Lower Church one finds the omnipresence of number three in different combinations, in the Upper Church, one finds similar arithmetic combinations of number five in dimensions (e.g. length 250 rf = 2x5x5x5) and in structural elaboration (e.g. five-lobate wall-piers). The number five cult and the ad quadratum design is largely visible in the Upper Church east façade, which could serve as the tangible synthesis of a rich and most interesting intangible program (**Figure 7g-h**).

4.6 ANDRIA (Italy), Castle del Monte, imperial castle, built 1241-1250

Maybe the best known and most enigmatic medieval building with regard to architectural design is the castle built mid XIII°century near the site of a former small Our Lady's convent in Andria (Puglia, Italy), commissioned by Frederic II (1194–1250), the then Holy Roman Emperor. It illustrates the amalgamation of European (Romanesque), classic antiquity (Roman) and Islamic design traditions, clearly combined with Christian semantics about life, rebirth, cosmic structure and the role of Jerusalem as religious and geographic center of the flat terrestrial world. Indeed, the octagon was since Babylonian times the main symbol for rebirth and eternal life (cfr. The supreme goddess Isthar was imaged by a octagonal star); the Pythagoreans loved the number eight for being the first prime 'regenerating' three times (8 = 2x2x2), and also the Christians used the octagonal plan for the baptistery, as this is the building where men get spiritual rebirth. In metaphysics, the number eight was the sum of three and five, two numbers with particular semantic meaning, and the octagon was associated with the person of Christ, as He was part of a tripartite God but also human being, using the five human senses (3 + 5 = 8), reborn the third day after death.

In the case of Castle del Monte, we propose a more tangible explanation for the triple octagonal design (i.e. the inner courtyard, the outside perimeter and eight octagonal edge-towers). As emerges from the plan and the geometry, the octagonal design seems a quasi-imitation, of the ca. 600 years earlier built Islamic octagonal 'Dome of the Rock' built on top of Mount Moriah in Jerusalem. This 'Dome of the Rock' as well as the adjacently located smaller 'Dome of the Chain' were at

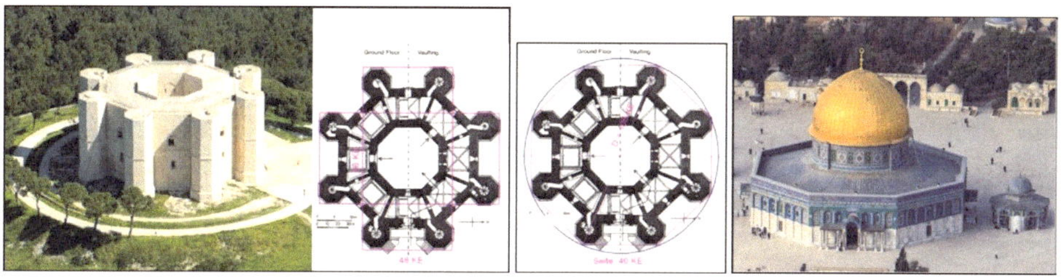

Figure 8.
Andria, Castel del Monte: (a) General aerial view. (b, c) squared and circle geometry of the ground floor (photo, drawings: Pinterest). (d) Jeruzalem, 'Dome of the Rock' and the adjacent, at east, small 'Dome of the Chain', both with octogonal plan,located in the center of the Temple Mount platform at Jeruzalem (Wikipedia).

Frederick's time converted into a Christian 'Templum Dominum', considered the geographic center of the world and also the most holy place for all three monotheist religions (Islam, Judaic and Christian). The 'bare stone platform', the center of the Dome of the Rock, was covered with the most extraordinary intangible content as it should have been the place where Abram sacrificed(Gen. 22), where Jacob dreamed (Gen. 28,11),where Solomon's Temple had his main altar (1Kings 6; 2 Cron.), and from where Mohammed ascended to heaven. Notwithstanding the Muslim origin and property of this Dome, it seems quite plausible that Frederick II, also crowned King of Jerusalem since 1229, was inspired by this unique symbolic content, and that he ordered the construction of an octagonal 'center of the world'- interpretation in his South Italian territories. Although the origin as a pagan (= Islamic) building, it was seen as inalienable part of the Christian heritage, and the imitation of similar octagonal building, substituting Solomon's Temple, was the best way to prove this towards his subjects. On top of this, it could be seen as a sign of obligingness from an illuminated and open minded Emperor towards the Muslim society (who had many Arabic scientists at his court) (**Figure 8d**).

There also might have been a second motive in play. Political history learns about the quasi permanent conflict between Frederick II and the then Pope Gregory IX. This last one had started in 1228 the before cited Sepulchral Basilica for S. Francis in Assisi, including also some small residence at the papal service, mainly for devotional reasons but also to consolidate papal political power in Central Italy. As the Castle del Monte was built in the same years (i.e. before 1240–1249) as the S. Francis Basilica at Assisi (built 1228–1253)), this Castle del Monte might also have been meant as a imperial answer in virtual confrontation with the papal project in Assisi. This hypothesis of ours can explain the evident link in the design with the biblical structure in Jerusalem and changes the Castle from a hunting refuge, as presented in literature, into a political and religious statement, expressed by this most unusual triple octagonal castle design. (One should note that the actual rectangular Al-aqsa mosque as well as the 'Dome of the Rock'- shrine has been damaged and rebuilt several times, but the mayor geometry of the 'Dome' has been preserved).

4.6.1 Geometry and numbers

Many hypothesis on the semantics of this exceptional format have been proposed, but still today, there are more questions than answers [24]. In most literature, it is presented as a 'hunting' castle, or part of a greater military defense chain of castles by Frederick II, but none of this hypothesis make sense in this totally isolated location on a ca. 500 m high top. As said, it seems more a political sign against Pope Gregory IX (building the S. Francis basilica on top of the Assisi's Collis

Inferiore), and a confirmation of the emperor's personal contribution as the leader of the 6th crusade, bringing the Jerusalem holy places under Christian control. The plausible intention of goodwill towards the Muslim world, the probable and significant astrologic input in the design of the castle, and the most probable sacral, social and political message gives this monument a complex and multilayered content, as intriguing as his design mixes the geometric and the arithmetic canons of Europe and the Mediterranean Basin.

The design draws two concentric octagons, with an open courtyard at the center and eight small octagonal towers at each corner of the external perimeter. Based on a possible (but not sure) metric unit of the 'imperial foot' of 0,52 m, the side of the overall circumscribed octagon signs (again) 40 feet and the ca.54,60 m diameter of the circle circumscription is to convert into the number 54,60: 0,52 = 105 feet, which indicates the particular arithmetic sequence of 105 = 1x3x5x7 i.e. the product of the first uneven prime numbers. The plan design is strictly squared in all particularities (**Figure 8b, c**). Each of the two floors is divided in eight trapezium-form rooms, covered by a central cross-rib square vault and two triangular side vaults. The parallel with the 'Dome of the Rock '-temple is evident as also this one is designed as a double concentric octagon signing the same overall diameter of 105 feet or ca. 55 m, what supports the hypothesis about a conscious imitation in Andria. A further connection with Islamic architecture is given by the slim-line corner-towers reminding the typology of the 4th century square towers alongside the byzantine defense wall round Constantinople, although the octagonal towers at the Andria-castle do not seem too much for defense as well as for semantic reasons.

However, any further comment on a probable metric modulation or other design parallels are hindered by sufficiently controllable historic and metric information about the initial composition (this last aspect hampered by the lack of net interior dimensions as all inside decorative marble wall-cladding has disappeared, what makes authentic reliable measuring impossible). Further research is needed.

5. Conclusion

The above mentioned small selection of architectural and structural characteristics illustrate the impact and extraordinary image building capacity of the ancient design procedures and the role of semantics and symbolisms. The Vitruvian maxims on a.o. ordinatio, analogia, symmetria, euritmia [1], as well as the Pythagorean semantics on quantity, harmonic numbering and arithmetic sequences were the guiding criteria in the historic design; this same criteria became universal quality indicators by which also modern designers can be inspired in the choice and the definition about the form and the geometry of their project. Recent E.U. research programs [25] have proved in several occasions the possible input of heritage building analysis by digital modeling; the application of similar techniques and algorithms might be reversible and be introduced in creative new design.

Sustainable architectural construction means not only an economic or utilitarian driven concern but implies a at least as important social and humanitarian assignment. The quality of contemporary projects, and the long-term guarantee for visual and functional quality in the built environment will certainly be enriched by a design which is not looking for eye-catching artistic effects, as this last ones change with the wind, but by a comprehensive design based on the universal human scaled modulation of all times, with a simply legible semantic communication.

Author details

André Frans De Naeyer
Faculty Design Sciences, Conservation Built Heritage, University Colleges Antwerpen, Belgium

*Address all correspondence to: andrefransdenaeyer@gmail.com

References

[1] GROS P., CORSO A., ROMANO E. 1997, *Vitruvio, De Architectura*, 2vol., Einaudi, Torino

[2] NORBERG-SCHULZ C.,1979, *Genius loci: towards a phenomenology of architecture*, New York

[3] KALVESMAKI J. 2013, *The theology of aritmetic number symbolism in platonism and early Christianity*, Washington DC

[4] GHYKA M.C. 1938, *Essai sur le rythme*, Paris

[5] DE BRUYNE E. 1946, Etudes d'esthétique médiévale, 3 vol., Bruges

[6] HAHNLOSER H.R. 1972 ^{2}, *Villard de Honnecourt, kritische Gesamtausgabe*, pp.403+Tafeln, Graz

[7] DE NAEYER A. (forthcoming), La Basilica di San Francesco in Assisi – La progettazione architettonica e il *design* medioevale, pp. 350 ca.

[8] UNESCO World Heritage List 1979, Memphis and its Necropolis – The Pyramid Fields from Giza to Dahshur – Description, Paris

[9] GHYKA M.C. 1931, *Le nombre d'or*, 2 vol.,Paris

[10] MICHETTI A.M. 2002, *Struttura e geometria nell'architettura storica*, in Zevi L., *Il manuale del restauro architettonico*, pp. Ù 73, Roma

[11] CONANT K.J. 1973^{3} , Carolingian and Romanesque Architecture 800-1200, Harmondsworth

[12] WINTERMANTEL S., 2017: *Geometrie, Maß und Zahl an der Aachener Marienkirche Karls des Großen,* in: Zeitschrift des Aachener Geschichtsvereins nrs.119/120, 2017-2018, pp. 51-194, Aachen

[13] AUBERT M., GOUBET S., 1958, *Cathédrales et trésors gothiques de France*, pp. 479, Paris

[14] CHARPENTIER L. 1977, *Les mystères de la cathédrale de Chartres*, pp.175, Ed. Lafont, Paris

[15]] FRANKL P., CROSSLEY P. 2000, *Gothic Architecture*, Yale

[16] JAMES J. 1977-82, *Chartres, les constructeurs*, 3 vol., Chartres

[17] VILLETTE J. 1978, *Le plan de la Cathédrale de Chartres – Hasard ou stricte géometrie*, Paris

[18] KOTTMANN A. 1981, *Fünftausend Jahre messen und bauen*, pp.296, Hoffmann Verlag, Stuttgart

[19] PADOVAN R.,1999, *Proportion, Science, Phylosophy*, London

[20] DE NAEYER A. 2019, The S. Francis Basilica at Assisi and the detection of medieval geometric design, ratio and modulation, in *International Journal of Architectural Heritage*, London; DOI 10.1080/15583058.2019.1683777

[21] DE NAEYER A. 2020, Numbers, geometry and modulation in the 13th cent. St. Francis Basilica at Assisi, in *Architettura Eremitica, Sistemi progettuali e paesaggi culturali,* Atti Convegno a cura di BERTOCCI S.& PARRINELLO S., Certosa del Galluzzo, pp. 120-125, Firenze, ISBN 978-88-09280-002-1

[22] ROCCHI G.C.d.Y. 1982, *La Basilica di S.Francesco ad Assisi – Interpretazioni e rilievo*, Firenze.

[23] MAGRO P. 2001, L'orientazione della chiesa doppia, in Basile G. & Magro P., *Il cantiere pittorico della Basilica di S.Francesco in Assisi*, Assisi

[24] SCHIRMER W. 1995 , *Castel del Monte: osservazioni sull'edificio*; in CALO MARIANI M.S, CASSANO R.(cura di): *Federico II, Immagine e potere*, Catalogo Mostra Bari 1995, pp.284-293, Marsilio Ed., Venezia

[25] E.U. Declaration of 09.04.2019 on *Cooperation on advancing digitalization of Cultural Heritage*, Brussel; and related projects, a.o. INCEPTION , 2020, E.U. Research Program on *Inclusive Cultural Heritage in Europe through time dynamic 3D semantic modeling,* University of Ferrara (I.).

Chapter 11

An Urban Paradox: Urban Resilience or Human Needs

Nilüfer Kart Aktaş and Nazlı Yıldız Dönmez

Abstract

Nowadays, metropolitan cities experience increasingly environmental problems as well as migration and urbanization pressure. As climate change, earthquake, flood, aridity and the last worldwide pandemic showed how cities are unprepared for these disasters. The ability of cities to cope with these disasters and survive depends on the existence and level of the city's resilience to these disasters. Also, the change and transformation of social structure effects the process of adaptation. Generally, urban citizens with economic power who have to live in crowded cities have created their own living areas in the periphery of cities with the desire to live away from the city and in nature. The population increasing every day due to migration from the city centers, attractiveness of natural life lead to urbanization of natural areas as well as the transformation of landscapes. The aim of this study is to measure the urbanization pressure, which is one of the important factors of landscape changes and to determine the results of the pressure for the important areas for resilience. In the scope this, it is detected the pressure of urbanization on the area and examined the landscape changes between the years of 2000-2020 in Istanbul/Zekeriyakoy. Zekeriyakoy, when it was a village until the 1980s, has been in the process of a radical change especially since 1987 and it has become an important center of attraction especially after the Marmara Earthquake in 1999. Corine Land cover and Google satellite data have been used to detect changes in the research. The main outcome of this study is; the district, which was dominated by agriculture and forest areas until the early 1990s, is now under intense pressure to settle and if the transformation occurs at the same speed, especially agricultural areas will almost disappear. This study is important in terms of how the field has changed in the years and the problems that this change will cause for the future. In this context it can be said that the change, transformation and adaptation expected to occur with the concept of urban resilience cannot be considered separate from human and human welfare.

Keywords: urban resilience, urban change, social change, spatial analysis

1. Introduction

Landscapes have affected by climate changes, land use changes and human-based complexities; and the mosaic structure within the landscape can change. All these changes can occured in different spatial dimensions and frequencies [1]. Changes in land use patterns play an important role especially in the growth of urban areas and in the transformation of land use patterns in rural or urban areas. In addition, important ecosystems such as agricultural areas, forests, coastal dunes and wetlands are the first and most adversely affected by this change [2–5]. However, these ecosystems are the most important areas that ensure the life cycle of cities and the resilience of cities.

Within used to create resistance against disaster such as flood, climate change and etc., the concept of resilience takes on the meanings of adaptation, change and transformation when it is used in relation to many different problems in the urban area.

UNISDR (2009) defines resilience as: "The ability of a system, community or society exposed to hazards to resist, absorb, accommodate to and recover from the effects of a hazard in a timely and efficient manner, including through the preservation and restoration of its essential basic structures and functions [6]."

Urban resilience consists on the capacity of a city and its urban systems to absorb the first damages, reduce the impacts arising from. Also, urban resilience, considering all kinds of disasters and disruptions in the systems of a city, is important to make communities more resilient when facing extreme events [7].

Urban resilience refers to the ability of an urban system-and all its constituent socio-ecological and socio-technical networks acrosstemporal and spatial scales-to maintain or rapidly return to desiredfunctions in the face of a disturbance, to adapt to change, and toquickly transform systems that limit current or future adaptiveca-pacity In this definition, urban resilience is dynamic and offers multiple pathways to resilience (e.g., persistence, transition, and trans-formation). It recognizes the importance of temporal scale, and advocates general adaptability rather than specific adaptedness. The urban system is conceptualized as complex and adaptive, andit is composed of socio-ecological and socio-technical networks that extend across multiple spatial scales [8]. As the concept of life has rich connotations and denotations that involve social, economic and ecological dimensions, along with multilevel interactions between human beings and the environment, the contradiction between supply and demand in daily life, as a collection of multiple pressures and even risks of socio-ecological systems, is an urgent problem of urban resilience [9].

Resilience in terms of cities generally refers to the ability to absorb, adapt and respond to changes in an urban system [10].

Urban systems' are conceptualized as complex, adaptive, emergent ecosystems composed of four subsystems; governance networks, networked material and energy flows, urban infrastructure and form, and socioeconomic dynamics that themselves are multi-scalar, networked and often strongly coupled [8].

The main common point of these definitions is that resilience is a way to improve a strategy/behavior to be able to survive and to adapt against external shifts/impacts. Basically, to construct resilience, the main ingredients are: resource, latitude (redundancy), networks (social and institutional), information, experience, knowledge, diversity and robustness [11].

With the concept of urban resilience, instead of returning to a stable balance point again, it would be more appropriate to talk about a new structure that understands and adapts to the change and transformation that occurs with different effects. In addition to the built environment, the harmony, learning, change and transformation of social structures stand out according to urban problems.

Resilience is not a characteristic that is evenly spread through the urban population. It depends crucially on the socially differentiated capacities of different groups and individuals. Poverty, gender, ethnicity and age have all been documented as contributing to differential vulnerability of social groups in cities to hazards like climate change, earthquake, flood, aridity, through features such as the quality of housing, location and access to services or social networks [12–15].

According to Biggs et al. [19], resilience as the capacity of a social-ecological system to sustain human well-being in the face of change, by persisting and adapting or transforming in response to change. A central challenge in this context is the capacity of social-ecological systems to continue providing key ecosystem services that underpin human well-being in the face of unexpected shocks as well as gradual, ongoing change [16–19].

The science of resilience helps deal with the uncertainties that arise from changes in land cover involving the interactions between environmental, social, and economic ecosystems within a city [20, 21].

In this context, it can be said that the change, transformation and adaptation expected to be achieved with the concept of urban resilience cannot be considered separate from human and human welfare. Although the human beings make the policies and their implementations to ensure urban resilience, the human causes the cities to be in this situation despite these policies. Then it is possible to ask the questions "resilience for whom?" and resillence or human needs.

2. Material and method

2.1 Material

2.1.1 Location of research area

The research area is located within the borders of Sarıyer in Istanbul. Zekeriyaköy, one of the oldest villages of Sarıyer is close to Maden, Bahçeköy, Uskumruköy, Demirciköy and Rumeli Feneri.

1.5% of the Zekeriyaköy-Uskumruköy Region is in the first degree archeological site, 5% is in the second degree archeological site and 93.5% is in the third archeological site.

Zekeriyaköy-Uskumruköy settlements are located in the north of the Istanbul and should be protected with their existing natural values. The region is in the interaction of forest and Bosphorus back view and it has a special position in terms of uses around it (**Figure 1**) [22].

Figure 1.
Location of Zekeriyaköy [23, 24].

2.1.2 Demographic structure

According to the 2019 Address Based Population Registration System (ABPRS) data, the population of Sarıyer District is 342,503 (https://www.nufusu.com/ilce/sariyer_istanbul-nufusu). Zekeriyaköy, which was a forest village until 2012, has started to be a neighborhood. The population in Zekeriyaköy is 18.707 in 2018 [25] (**Table 1**).

There is a rapid increase in the population development of the settlements in the adjacent area of Sarıyer District between 1995 and 1997 and 1997-2000 (**Figure 2**). Especially after the earthquake of 17 August 1999 (due to the solid ground of the settlement, infrastructure and ease of transportation), it has been used as the first residential area by the upper income group.

2.1.3 Socio-cultural structure

As a result of the great wave of immigration caused by the war during the 93 War, several families from Caucasus and Crimea were settled in Zekeriyaköy. In addition to this, migrations have been received from the Black Sea Region [26].

Until the 1980s, it looked like a village and recreation area consisting of 70 houses with 2 floors and a garden but especially since 1987, it has gone through a period of radical change. While Zekeriyakoy began to change rapidly since the 1990s, it became an important center of attraction as a result of the solid ground of the Sarıyer region and the naturalness of the village, especially after the Marmara Earthquake of 1999. The area has turned into a region close to the city center because of the high income group who prefer a quiet environment and expensive villas [22].

In Zekeriyakoy, old residents and new comers live together. These different groups create a multicultural but less interactive social structure. New arrivals in the region are representatives of a different culture with their educational level, professional and economic accumulation, while the inhabitants of the old villages are people engaged in animal husbandry and agriculture, producing according to feudal society characteristics [27].

2.2 Method

In this study, it is aimed to reveal the landscape changes between 2000 and 2019 in Zekeriyakoy. Corine Land cover and Google satellite data have been used to detect changes in the research. Using the Quantum GIS program for transportation lines, the transportation routes have been taken from the data provided by Open Street Map and transferred to ArcGIS environment.

Corine Land Cover data have been used to determine the land changes for the years 2000 - 2006 – 2012 – 2018 and 2019. Corine (Coordination of Information on the Environment) is the land cover/use data generated by computer aided visual interpretation method based on satellite imagery according to the Land Cover/Use Classification set by the European Environment Agency. In the Corine Land Cover, the artificial zones are classified as urban spot areas in the study under the categories 111 and 112 under the urban structure classification and the construction sites under code 133 under the artificial sites. The areas coded with 211 and 222 in the agricultural areas in the section of Corine land cover have been classified as agricultural areas within the scope of the study and the areas coded with 311, 312, 333 in the forest and semi-natural areas have been classified as forest areas. In 2015, Google satellite photo ArcGIS 10 program coordinates the agricultural areas, forest areas and urban stain areas manually. The maps obtained between 2000 and 2019 have overlapped and Spatial Difference and Spatial Intersection analyzes have been performed in GIS environment and physical changes of landscapes have

DOI: http://dx.doi.org/10.5772/intechopen.95271

Year	2000	2007	2008	2009	2010	2011	2012	2013	2014	2015	2016	2017	2018	2019
Population	7323	8230	11,279	12,528	13,817	14,755	15,750	16,524	17,581	18,457	18,833	18,867	18,707	19,896

Table 1.
Zekeriyaköy Population for the last 19 years.

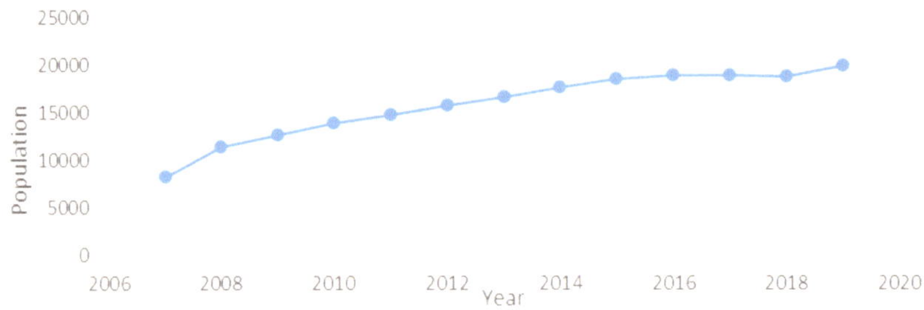

Figure 2.
Population of Zekeriyaköy by Years.

been determined quantitatively and graphs have been created. Density analysis was performed by determining the density of the buildings according to years with the Spatial Analysis Tools - Kender Density in the ARCGIS program.

3. Findings

Within the scope of this study, the land use data for Zekeriyaköy for the years 2000-2019 have been generated. In this study conducted to determine the changes between 2000, 2006, 2012, 2018 and 2019, it is seen that forest areas occupy the largest area.

According to the study in 2000, urban stain areas are 330.6 ha, agricultural areas are 161.4 ha and forest areas are 587.6 ha (**Figures 3** and **4**). According to this, forest areas constitute the highest area with 54% in the spatial distribution, while urban areas constitute 31% and agricultural areas constitute 15% (**Figure 4**).

According to the study in 2006, forest areas are 558.4 ha, urban strain areas are 378.1 ha and agricultural areas are 143.1 ha (**Figures 5** and **6**). According to this, forest areas constitute the highest area with 52% in the spatial distribution, while urban areas constitute 35% and agricultural areas constitute 13%.

According to the study in 2012, urban areas are 423.4 ha, agricultural areas are 97.7 ha and forest areas are 558.5 ha (**Figures 7** and **8**).

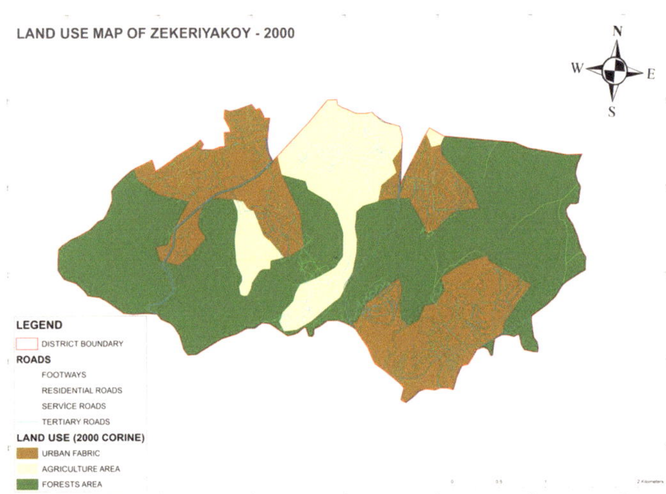

Figure 3.
Land use map in 2000.

According to this, forest areas constitute the highest area with 52% in the spatial distribution, while urban areas constitute 39% and agricultural areas constitute 9%.

According to study in 2018, urban areas are 456 ha, agricultural areas are 97.7 have forest areas are 525.9 ha (**Figures 9** and **10**). According to this, forest areas constitute the highest area with %48,7 in the spatial distribution, while urban areas constitute % 42,23 and agricultural areas constitute %9,05.

According to the study in 2019, urban areas are 478.1 ha, agricultural areas are 72 ha and forest areas are 529.5 ha (**Figures 11** and **12**).

According to this, forest areas constitute the highest area with 49% in the spatial distribution, while urban areas constitute 44% and agricultural areas constitute 7%.

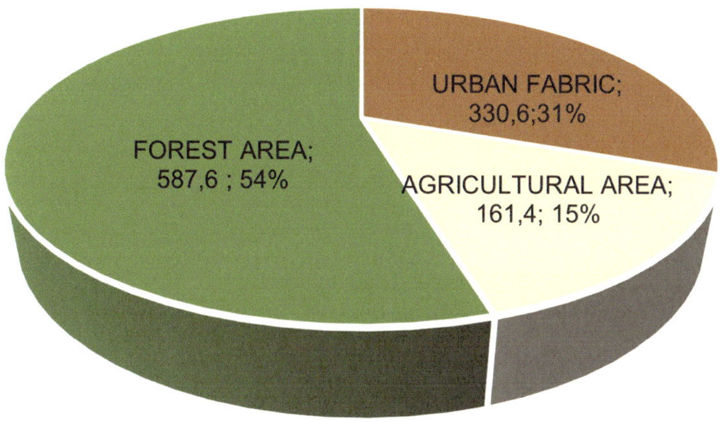

Figure 4.
Distribution of land use in 2000.

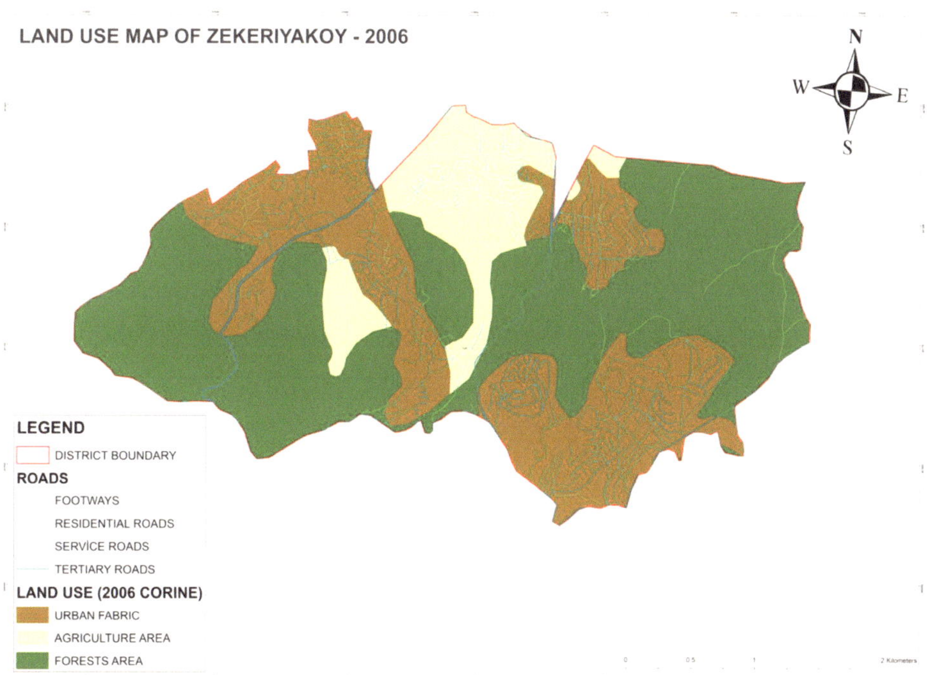

Figure 5.
Land use map in 2006.

4. Conclusion

The changes and regional spreads over the years enable projections about the future of that region. Analyzing, understanding and managing the changes both physical and social are necessary in order to protect and ensure the continuity of areas such as wetlands, agricultural areas, forest areas that have special importance in ensuring urban resilience.

Determination of landscape change, it provides the relationship between landscape structure and process. Any change that will occur will affect the landscape structure and will lead to changes in living environments and to a large extent shrink.

Zekeriyakoy, when it was a village until the 1980s, has been in the process of a radical change especially since 1987 and it has become an important center of

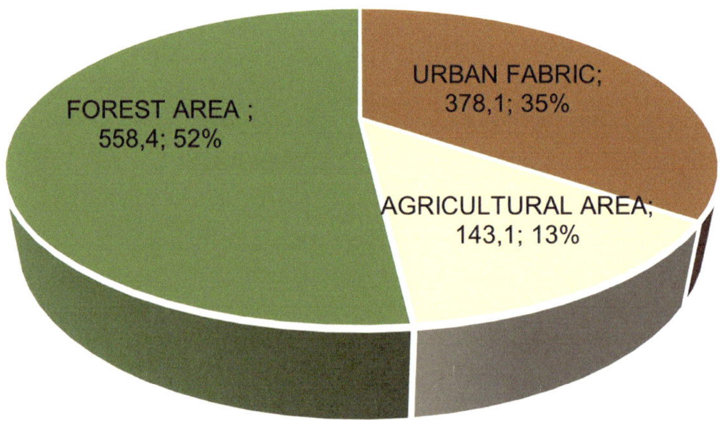

Figure 6.
Distribution of land use in 2006.

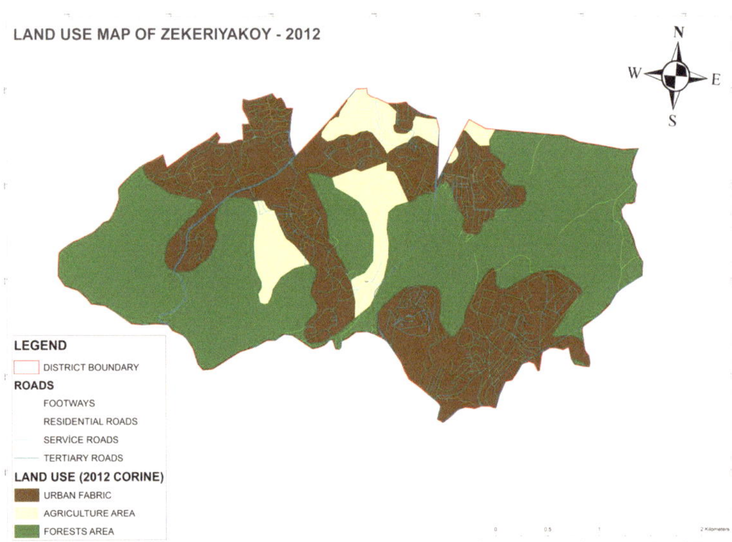

Figure 7.
Land use map in 2012.

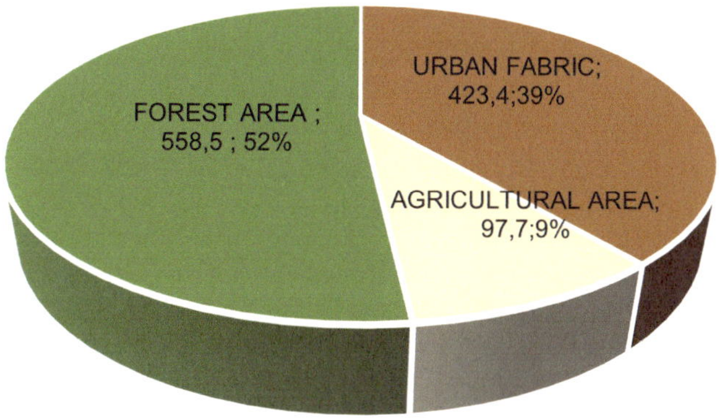

Figure 8.
Distribution of land use in 2012.

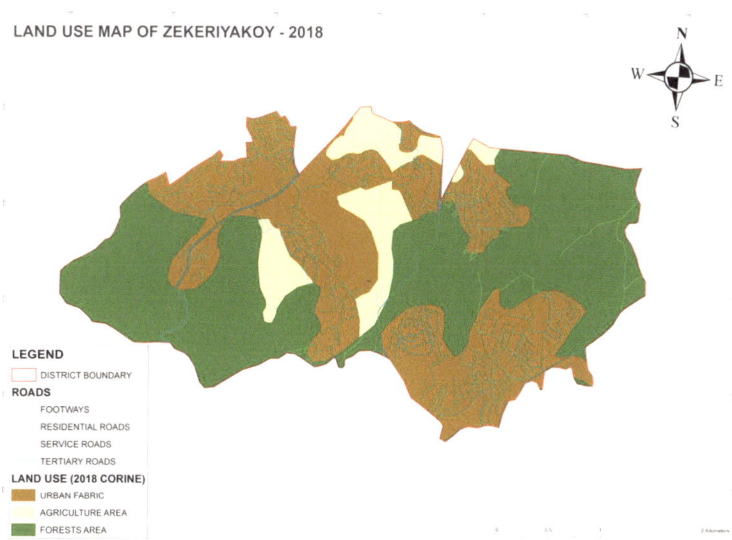

Figure 9.
Land use map in 2018.

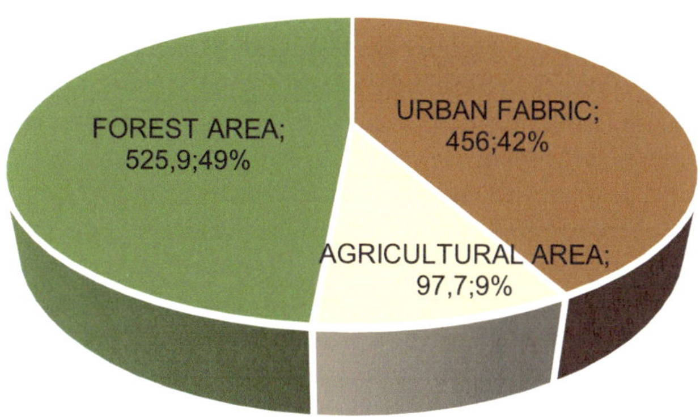

Figure 10.
Distribution of land use in 2018.

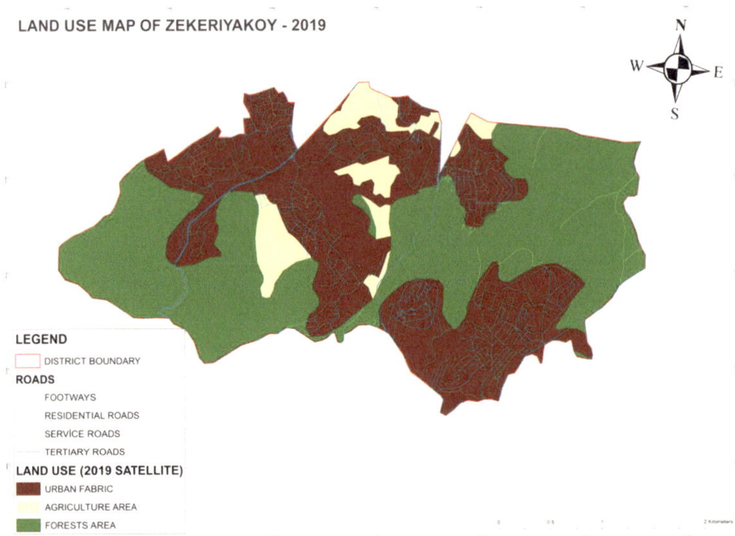

Figure 11.
Land use map in 2019.

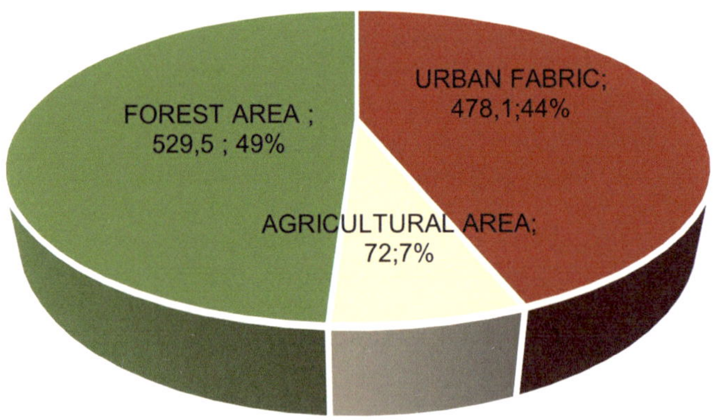

Figure 12.
Distribution of land use in 2019.

attraction especially after the Marmara Earthquake in 1999. Together with the physical change, the greatest change has been experienced in the social structure. The area has become a region where there are sites with expensive villas and preferred by the residents in high-income group who prefer close to the city center and a quiet environment. The earthquake in 1999 played a major role in this change, which has been observed since 2000. Single-storey or 2-storey villas and sites on solid ground have become an attractive living space for high-income groups. Thus, there has been a major change in the socio-economic structure of Zekeriyaköy.

Zekeriyaköy, which entered the rapid urbanization process due to its proximity to the city center and ease of transportation, suffered significant losses in forest and agricultural lands during this period. In particular, urban stains such as residential areas were created instead of agricultural land, and thus a large part of fertile agricultural land became unusable. In approximately 20 years, 89.4 hectares of agriculture and 58.1 hectares of forest area have been lost. The majority of these areas have turned into independent villas and closed sites.

There is a significant amount of change between the years 2000-2019. It is observed that forest areas remain the largest area in this 20-year period of change, while serious reductions are experienced (**Table 2**).

Between 2000 and 2019, the greatest spatial change in landscape changes in Zekeriyaköy occurred in urban stain areas. Urban stain areas increased by 47.5 ha between 2000 and 2016 and the change rate was 14.37%. Between 2006 and 2012, it increased by 45,3 ha and the change rate was 11,98%. Between 2012 and 2018, urban strain area is 32,6 ha and the change rate is 7,69%. Between 2018 and 2019, urban strain area is 22,1 ha and the change rate is 4,85%.

Between 2000 and 2012, agricultural land decreased. The majority of urban stains were spread in agricultural areas. The greatest change in the agricultural fields was experienced between 2006 and 2012. Between 2006 and 2012, agricultural land decreased as 45.4 ha (**Table 3**) (**Figure 13**).

As a result of the studies, urban stains have increased in the last 19 years in Zekeriyaköy as 147.5 ha. 89.4 hectares of this increase was due to the decrease of 58.1 hectares of forest area.

In addition, it is observed that this serious increase in housing problem is very intense in certain regions, but there is a serious risk of spreading to the whole region (**Figure 14**).

This study is important in terms of how the field has changed in 20 years and the problems that this change will cause for the future. It is possible to control the urbanization pressure on the villages close to cities and to protect the identity of the villages or to protect them together with natural landscape areas and to plan them with an integrated approach.

The importance of forest areas and agricultural areas was further understood during the Covid 19 pandemic process. These areas will ensure the resistance of cities to disasters. The study area became an escape area after a major earthquake in Istanbul. However, if the development and population increase continue in the region, the loss of forest and agricultural areas will continue rapidly.

One of the biggest problems of today is the increasing urban population and it has spread metropolises and the surrounding villages, rural areas and cities. Although it is known that agriculture is the struggle to save our future today, agricultural areas are rapidly being transformed into structured areas.

It is necessary to look for answers to questions about who is responsible for this situation or how it can be prevented. Of course, whoever is responsible are not urban people who moved to a safer area after a disaster. The main problem is that the urban does not have a functioning plan and cannot be protected. Nowadays, pandemic and disasters caused by climate change require the preparation and implementation of an emergency action plan. In addition, citizens should be informed about the importance of natural resources such as agricultural areas,

Years	Urban fabric (ha)	Agriculturel area (ha)	Forests area (ha)	Total (ha)
2000	330,6	161,4	587,6	1079,6
2006	378,1	143,1	558,4	1079,6
2012	423,4	97,7	558,5	1079,6
2018	456	97,7	525,9	1079,6
2019	478,1	72	529,5	1079,6

Table 2.
Land cover amounts between 2000 and 2019.

Class	Years	Amount of change	Increase/ decrease	Change rate
Urban fabric	2000-2006	47,5	↑	14,37%
	2006-2012	45,3	↑	11,98%
	2012-2018	32,6	↑	7,69%
	2018-2019	22,1	↑	4,85%
Agricultural area	2000-2006	18,3	↓	11,33%
	2006-2012	45,4	↓	31,73%
	2012-2018	0	—	0%
	2018-2019	25,7	↓	26,31%
Forest area	2000-2006	29,2	↓	4,97%
	2006-2012	0,1	↑	0,01%
	2012-2018	32,6	↓	5,19%
	2018-2019	0	—	0%

Table 3.
Area change rates between 2000 and 2019.

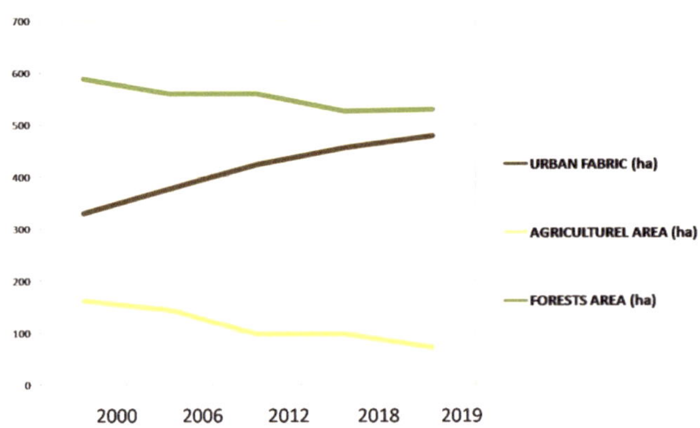

Figure 13.
Distribution of land uses of Zekeriyaköy.

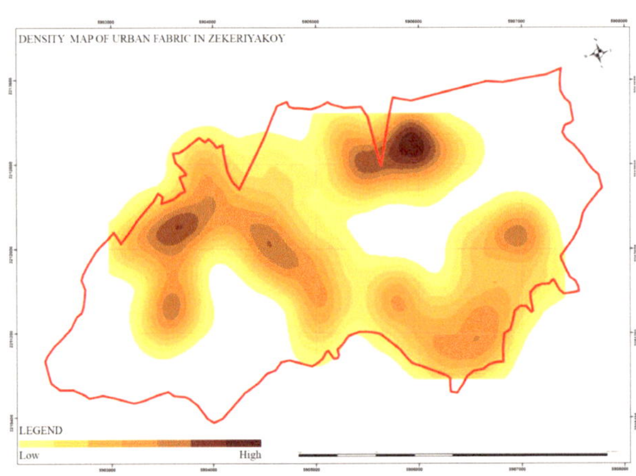

Figure 14.
Distribution of density of urban fabric in Zekeriyaköy in the last 19 years.

DOI: http://dx.doi.org/10.5772/intechopen.95271

forest areas and water resources. The future of the city and its citizen depends on the protection of natural resources and this conservation awareness should be created. It should not be forgotten that the most important need of human is shelter, nutriment and water. This will only be possible by preserving the balance of nature and natural resources.

Author details

Nilüfer Kart Aktaş[1]* and Nazlı Yıldız Dönmez[2]

1 Department of Landscape Architecture, Istanbul University Cerrahpasa, Istanbul, Turkey

2 Msc. Landscape Architect, Istanbul, Turkey

*Address all correspondence to: niluferk@istanbul.edu.tr

References

[1] Farina, A. Landscape Ecology in Action. Kluwer Academic Publishers, Netherlands; 2000.

[2] Jenerette, G. D. and Wu, J. Analysis and Simulation of Land-Use Change In The Central Arizona – Phoenix Region, USA. Landscape Ecology, 16; 2001. p 611-626.

[3] Erdem, U. ve Doygun, N., Toprak Potansiyeline Uygunluğu Yönünden Bornova İlçesi Alan Kullanım Yapısının İncelenmesi. Ekoloji 2011 Sempozyumu, 5-7 Mayıs, Düzce; 2011. P. 135.

[4] Mendoza, M. E., Granados, E. L., Geneletti, D., Perez-Salicrup, D. R. ve Salinas, V. Analysing Land Cover and Land Use Change Processes at Watershed Level: A Multitemporal Study in the Lake Cuitzeo Watershed, Mexico (1975-2003). Applied Geography, 31;2011. P.237-250.

[5] Doygun, N. Determining Interactions Between Land Use Potential And Land Use Structure In Bornova District [thesis].Ege University Graduate School of Natural and Applied Science Landscape Architecture Department; 2012.

[6] UNISDR 2009, Terminology on Disaster Risk Reduction. http://www.unisdr.org/eng/library/UNISDR-terminology-2009-eng.pdf.

[7] Ribeiro P. J. G., Gonçalves L. A. P. J. Urban resilience: A conceptual framework. Sustainable Cities and Society 2019 50.

[8] Meerow S., Newell, J. P. and Stults, M.Defining urban resilience: A review. Landscape and Urban Planning 147 (2016) p.38-49.

[9] Sun, H., Zhen, F., Lobsang, T., Li, Z. Spatial characterisrics of urban life resilience from the perspective of supply and deman: A case study of Nanjing, China. Habitat İnternational 88 (2019).

[10] Desouza, K. C. and Flanery, T. H. Designing, Planning, and Managing Resilient Cities: A Conceptual Framework. Cities; 2013 p.35, p89-99. http://dx.doi.org/10.1016/j.cities.2013.06.003.

[11] Kundak, S. Radix of Resilience. Journal of Resilience, 1(1); 2017 p.55-69, ISSN: 2602-4667.

[12] Tyler, S., Moench, M. A framework for urban climate resilience. Climate and Development;2012. P.4:4, 311-326, DOI: 10.1080/17565529.2012.745389

[13] Moser, C., & Satterthwaite, D.Towards pro-poor adaptation to climate change in the urban centers of low- and middle-income countries. In R. Mearns & A. Norton (Eds.), Social dimensions of climate change: Equity and vulnerabiity in a warming world Washington, DC: World Bank;2010 p. 231– 258.

[14] Pelling, M. The vulnerability of cities: Natural disasters and social resilience. London: Earthscan; 2003.

[15] Satterthwaite, D., Dodman, D., & Bicknell, J. Conclusions: Local development and adaptation. In J. Bicknell, D. Dodman, & D. Satterthwaite (Eds.), Adapting cities to climate change: Understanding and addressing the development challenges. London: Earthscan; 2009 p. 359-383

[16] Folke, 2016. Resilience. Ecology and Society 21(4):44. https://doi.org/10.5751/ES-09088-210444.

[17] Bennett, E. M., Peterson, G. D. and Gordon, L. J. Understanding Relationships Among Multiple Ecosystem Services. Ecology Letters; 2009. 12:1394-1404. http://dx.doi.org/10.1111/j.1461-0248.2009.01387.x.

[18] Carpenter, S. R., Mooney, H. A., Agard, J., Capistrano, D., DeFries, R. S., Diaz, S., Dietz, T., Duraiappah, A. K., Oteng-Yeboah, A., Pereira, H. M., Perrings, C., Reid, W. V., Sarukhan, J., Scholes, R. J. and Whyte, A. Science for managing ecosystem services: beyond the Millennium Ecosystem Assessment. Proceedings of the National Academy of Sciences, USA; 2009. p. 106:1305-1312. http://dx. doi.org/10.1073/pnas.0808772106.

[19] Biggs, R., Schlüter M. and. Schoon, M. L. Principles for Building Resilience: Sustaining Ecosystem Services in Social-Ecological Systems. Cambridge University Press, Cambridge, UK; 2015 http://dx.doi.org/10.1017/cbo9781316014240.

[20] Wang, S.H., Huang, S.L., Budd, W.W. Resilience analysis of the interaction of between typhoons and land use change. Landscape and Urban Planning 2012 106(4) p. 303-315.

[21] Du, M., Zhang, X., Wang, Y., Tao, L., Li., H. An operationalizing model for measuring urban resilience on land expansion. Habitat International 2020 (102).

[22] Anonymous 1. 1/5000Scale Zekeriyaköy and Uskumruköy Area Conservation Development Plan Report, Sarıyer Municipality Directorate of Zoning and Urban Planning, İstanbul.

[23] http://www.wikizero.biz/index.php?q=aHR0cHM6Ly90ci53aWtpcGVkaWEub3JnL3dpa2kvQmFow6dla8O2eSxfU2FyxLF5ZXI. Retrieved: 03.06.2019.

[24] https://www.google.com.tr/maps/place/Zekeriyaköy,+Zekeriyaköy+Mahallesi,+Sarıyer%2Fİstanbul/@41.2010283,28.9932499,13z/data=!3m1!4b1!4m5!3m4!1s0x409fe054c7e97837:0x604e2d7504a9b996!8m2!3d41.1992!4d29.0284

[25] https://biruni.tuik.gov.tr/medas/?kn=95&locale=tr

[26] http://www.wikizero.biz/index.php?q=aHR0cHM6Ly90ci53aWtpcGVkaWEub3JnL3dpa2kvWmVrZXJpeWFrw7Z5LF9TYXLEsXllcg (Retrieved: 10.06.2019).

[27] Güler, M. Zekeriyaköy'ün Eski ve Yeni Yüzü (Kentlerde Yeni Yerleşim Alanları ve Uçlara Kayma), VI. Ulusal Sosyoloji Kongresi, "Toplumsal Dönüşümler ve Sosyolojik Yaklaşımlar", Adnan Menderes Üniversitesi, Aydın; 2009.